D1879555

Molecular Approaches in Plant Improvement

Molecular Approaches in Plant Improvement

Dr Md Aminul Islam
Dr Priyanka Dhakate

KALPAZ

©Author

All rights reserved. No part of this work may be reproduced, stored, adapted, or transmitted in any form or by any means, electronic, mechanical, photocopying, micro-filming recording or otherwise, or translated in any language, without the prior written permission of the copyright owner and the publisher. The book is sold subject to the condition that it shall not, by way of trade or otherwise, be lent, resold, hired out, or otherwise circulated without the prior publisher's written consent in any form of binding or cover other than that in which it is published.

The views and opinions expressed in this book are author(s) own and the facts reported by them have been verified to the extent possible, and the publishers are not in any way liable for the same.

First Published, 2018

Published by

Kalpaz Publications
C-30, Satyawati Nagar,
Delhi – 110052
E-mail: kalpaz@hotmail.com
Ph.: 9212142040

Printed at: G. Print Process, Delhi

Cataloging in Publication Data—DK
Courtesy: D.K. Agencies (P) Ltd. <docinfo@dkagencies.com>

Islam, Md Aminul *(Post-doctoral research fellow)*, **author.**
Molecular approaches in plant improvement / Dr. Md Aminul Islam, Dr. Priyanka Dhakate.
pages cm
Includes bibliographical references.
ISBN 9789386397829 (HB)

1. Growth (Plants)—Molecular aspects. 2. Plant molecular genetics. I. Dhakate, Priyanka, author. II. Title.

LCC QK731.I85 2018 | DDC 581.3 23

Contents

List of Contributors

Suneel Kumar: National Bureau of Plant Genetic Resources, DPS Marg, New Delhi 110012, India.

Uttam Kumar: Borlaug Institute for South Asia (BISA), PAU Campus, Ludhiana, Punjab 141004 India.

Arun Kumar Joshi: The International Maize and Wheat Improvement Center (CIMMYT), Borlaug Institute for South Asia (BISA), G-2, B-Block, NASC Complex, DPS Marg, New Delhi 110012, India.

Md Aminul Islam: National Institute of Plant Genome Research, Aruna Asaf Ali Marg, New Delhi-67.

Shashi Bhushan Tripathi: The Energy and Research Institute, India Habitat Centre, Lodhi Road, New Delhi-03.

Priyanka Dhakate: TERI University, 10 Insitutional Area, Vasant Kunj New Delhi 10075.

S.M. Shivaraj: Departement de Phytologie, University Laval, Quebec, QC, Canada.

Anandita Singh: TERI University, 10 Insitutional Area, Vasant Kunj New Delhi 10075.

Anirban Chakraborty, Subodh Verma and Sabhyata Bhatia: National Institute of Plant Genome Research, Aruna Asaf Ali Marg, New Delhi 110067.

Pratima Pandey and Shashi Bhushan Tripathi: The Energy and Research Institute, India Habitat Centre, Lodhi Road, New Delhi-03.

Santosh Kumar and Sabhyata Bhatia: National Institute of Plant Genome Research, Aruna Asaf Ali Marg, New Delhi 110067.

Baljinder Singh, Manish Tiwari and Sabhyata Bhatia: National Institute of Plant Genome Research, Aruna Asaf Ali Marg, New Delhi 110067.

Sahil Mehta: International Centre For Genetic Engineering And Biotechnology, Aruna Asaf Ali Marg, New Delhi 110067.

Shyam Sunder Sharma, Shashi Bhushan Tripathi: The Energy and Research Institute, India Habitat Centre, Lodhi Road, New Delhi-03.

Preface

Present century is slated to be the age of life sciences, given its relevance to human survival. Agricultural research has evolved over a century from traditional human selection to use of molecular markers to establish current crops. Traditional forward genetic approaches have proved instrumental in selection and establishment of traits in innumerable crop improvement programs across the globe. In 1980's introduction of reverse genetic approaches revolutionized the then established crop improvement programs and ably aided them towards a brighter future. More recently with advent of next generation sequencing platforms crop sciences research has seen another paradigm shift. Reverse genetic approaches have now amalgamated with forward genetic approaches to yield faster manipulation of crops.

This book presents collation of modern day approaches and methods that promise to yield results on field to aid in crop improvement. This book shall prove helpful in understanding current trends in crop sciences in era of next generation sequencing platforms.

Md Aminul Islam
Priyanka Dhakate

Acknowledgements

It we are grateful to Dr. Sabhyata Bhatia, Scientist VI, National Institute of Plant Genome and Research for her support and encouragement. We acknowledge Dr. Shashi Bhushan Tripathi, Dr. Anandita Singh, TERI University in compiling the book and being a constant source of encouragement. We acknowledge the contributors of all the papers for their diligence and efforts to revise and review the chapters. We acknowledge Gyan book publishers for accepting and publishing book.

Thanking you

Md. Aminul Islam

Priyanka Dhakate

Chapter - 1

Spot Blotch Disease of Wheat: Present, Past and Future

—*Suneel Kumar, Uttam Kumar, Arun Kumar Joshi*

1.1 Introduction

The world's population is increasing by one billion in every 11 years and with this rate, it is expected to be 8.5 billion by the year 2025. Wheat is widely grown staple food crop of nearly 36 per cent of the world's population, occupying 17 per cent of crop acreage worldwide. In terms of harvested area, wheat (mainly common/soft wheat *Triticum aestivum,* but also durum/hard wheat *T. turgidum*) is the largest crop worldwide (about 217 million hectares in 2010; FAO 2012) and ranks third in the global annual production of commodities (about 651 million tons in 2010; FAO 2012). The world's main wheat producing countries are China, India, United States, Russian Federation, France, Australia, Germany, Ukraine, Canada, Turkey, Pakistan, Argentina, Kazakhstan and United Kingdom. Majority of cultivated wheat varieties at present belongs to hexaploid wheat *(Triticum aestivum L.),* which is known as common bread wheat and valued for bread making. India ranks the second largest wheat-producing nation in the world and contributes approximately 11.7 per cent to the world wheat production from 12 per cent global area. China is the largest producer of wheat with 11 per cent and 15 per cent contribution respectively, in area and production. In India, wheat is being consumed by more than 65 per cent of population and is likely to increase further due to change in food habits. Wheat is mostly consumed in the form of chapatti for which wheat is cultivated on approximately 95 per cent of arable land in India. Wheat suffers from several diseases and development of disease resistant genotypes is the best solution to manage wheat diseases. Wheat breeders have been successful in developing genotypes that are resistant to some of the diseases (Duveiller and Sharma 2009).

Among the wheat disease, spot blotch is dominant and one of the most important diseases of wheat (*T. aestivum* L. em Thell) in warmer and humid regions. Spot blotch disease is caused by *Bipolaris sorokiniana* (Sacc.) Shoem. syn. *Drechslera sorokiniana* (Sacc.) Subrm and Jain (syn. *Helminthosporium sativum*, teleomorph *Cochliobolus sativus*) and not only confined to certain parts of India but it occurs in several parts of the world particularly in South East Asia and Latin American countries where nearly 12 million hectares of land under cultivation are affected (Nagarajan and Kumar 1998; Joshi et al. 2007b). The specific area where it is most prevalent includes African countries (Tanzania, Zambia and Madagascar), Asia (High rainfall areas of China, India, Bangladesh, Nepal, Indonesia, Thailand and Philippines), Latin America (North East Argentina, Brazil, Bolivia and Paraguay) and states of America (Indiana, Kansas, Minnesota, Montana, North Dakota, South Dakota, Utah, Virginia). In USSR, Northern Kazakhstan is reported a zone of the highest severity of Spot Blotch (Khasanov 1988; Villareal et al. 1995).

Spot blotch was first time reported by Mohy in 1914 (Joshi 1986) but attention was paid only after green revolution (Saari 1998) as the production of wheat has expended in newer growing areas characterized by high temperature (coolest month greater than 17^{0}C) and high humidity (van Ginkel and Rajaram 1993). Ideal conditions for spot blotch development on the leaves are relative humidity near to 100 per cent caused by rainfall, irrigation or dew with temperatures 20^{0}C to 30^{0}C and long day periods (>12 to 18 h). Increase in temperature coupled with high relative humidity and soil nutrient condition favours the outbreak of the disease (Ragiba et al. 2004). Disease intensity is directly proportional to rainfall under hot weather condition. Other factors which increase foliar blight intensity are minimum tillage, frequent irrigation, late planting and inappropriate fertilizer regimes. The disease becomes severe especially during grain filling stage (Joshi and Chand 2002) and causes significant grain yield losses up to 100 per cent under most severe conditions (Sharma et al. 1997a). The pathogen affects almost all the crops belonging to Gramineae family, though the chances of migration of isolate of one crop to other crop are rare (Bakonyi et al. 1998; Pandey et al. 2005). *B. sorokiniana* also causes root rot, seed rot, crown rot, black point on grains, head blight, hamper germination and seedling blight of wheat grains (Zillinsky 1983; Kumar et al. 2002). *B. Sorokiniana* is both, seed and soil borne disease and transmitted by seed themselves in nature (Neergaard 1997). Seed infection by *B. sorokiniana* can adversely affect germination and root system development and can kill seedlings within a few days (Mehta 1993). The fungus *Cochliobolus sativus* is the teleomorph (sexual stage) of *B. Sorokiniana* (anamorph) which is the causal agent of wide variety of diseases.

There have been several discussions among the scientists whether the spot blotch caused in barley (*Hordeum vulgare*) is same or different than the one cause spot blotch in wheat (*Triticum aestivum*). It was concluded that wheat spot blotch and barley spot blotch are caused by same pathogen i.e., *Bipolaris sorokiniana* but the pathotypes are different. Chand et al (2003) studied the migration of barley leaf blight pathogen to wheat and vice versa. It was found that the pathogen from barley is not able to infect wheat plant and the pathogen from wheat is unable to infect barley. It proves that the two types of pathotypes may not be same at least for pathogenicity. However, other characteristics are same in term of life cycle, mode of infection, reproduction, morphology, growth habits and even symptoms on the host plant. Mining of literature revealed that the same pathogen name (*Bipolaris sorokiniana*) was used for wheat spot blotch and barley spot blotch. The pathogen causing spot blotch in wheat and barley are classified in the same manner (table 1).

Previously, *B. sorokiniana* was classified as; division Eumycota, subdivision Deuteromycotina, class Hyphomycetes, subclass Sporomycetidae, order Moniliales and family Dematiaceae (Shoemaker 1959) but after the discovery of perfect stage in *B. sorokiniana* (Raemaekers 1988), the fungus was transferred under the subdivision Ascomycotina, class Loculoascomycetes, order Pleosporales and family Pleosporaceae (Table 1). This pathogen is capable of causing damage starting from the seedling and primary leaf stage, though the plant tends to become more susceptible after flowering but greatest damage occurs during the flowering to milky stage of wheat plant. All of the plant parts including roots, leaves, nodes, spikes and grains are attached by *Bipolaris*. The small spots on the leaf, if numerous, coalesce and the leaf prematurely dries up, reducing the photosynthetic area of the plant.

Table1: Classification of pathogen *Bipolaris*

	Old classification	***New classification***
Division	Eumycota	Eumycota
Subdivision	Deuteromycotina	Ascomycotina
Class	Hyphomycetes	Loculoascomycetes
Subclass	Sporomycetidae	-
Order	Moniliales	Pleosporales
Family	Dematiaceae	Pleosporaceae

The sexual state of *Bipolaris* is heterothallic and each thallus is bisexual and self-sterile (Tinline 1951). The sexual state of *Bipolaris* is not common in nature but has been produced under laboratory conditions. Environmental conditions, such as substrate, temperature and pH influence perithecia formation (Tinline and Dickson 1958). Conidia of *Bipolaris* can be white to olivaceous but this is not linked to mating type (Tinline and Dickson 1958). *Bipolaris* is highly variable in pathogenicity and virulence. Kline and Nelson (1963) identified 13 pathogenicity types from 28 isolates based on their varying capacity to cause disease on selected species of Poaceae while Zhong and Steffenson (2001) identified two pathotypes of *Bipolaris* from the USA.

1.2 Yield loss

Yield losses due to spot blotch are reported to vary from 15.5 to 19.6 per cent (Dubin and Van Ginkel 1991), 20 to 80 per cent (Duveiller and Gilchrist 1994), and may be up to 100 per cent under severe conditions (Mehta 1994). Globally, an estimated 25 million hectares (million hactor) of wheat cultivated land is affected by spot blotch disease (van Ginkel and Rajaram 1998). It has been estimated that infected plots yields nearly 43 per cent less as compared to the fungicide protected (Villareal et al. 1995). Indian subcontinent has 10 mha of affected land, out of which India alone has 9 mha, most of which is in the rice wheat cropping system (Nagarajan and Kumar 1998). In Bangladesh, 15 per cent wheat yield losses, on average, were attributed to spot blotch infection for the wheat crop season 1991-1994 (Alam et al. 1994). In trials on 13 farmers' fields in 1991, losses in variety UP262 were around 10.7 per cent. However, in an experiment at the Rupandehi research station in the same year using the same variety, 42 per cent disease infection at late crop stages were recorded, causing a 25 per cent yield loss (Dubin and Bimb 1994). In similar trials in previous years, infection levels were 47 per cent and 88 per cent, resulting in yield losses of 13 per cent and 27 per cent, respectively. Yield loss estimates due to spot blotch on wheat vary widely. For example, 85 per cent losses were reported from Zambia and 40 per cent from the field trials in Philippines. In addition, yield losses with the highly susceptible cultivar 'Mitacore' in an experiment conducted at Londrina, Brazil ranged from 79 to 87 per cent, and the disease severely affected grain quality. Millano and Zadoks (1985) found a 38 per cent yield loss using African wheat cultivars in growth-chamber studies in the Netherlands. Seed infection reduces seedling emergence by up to 38 per cent (Clark and Wallen 1969).

An independent study was conducted by Sharma et al. (2002) for two years to determine the reduction in grain yield and its components in a newly released Nepali wheat cultivar, Gautam, on resource-poor farms under a rice–wheat cropping sequence. The disease severity was recorded up to 100 per cent and 70 per cent and yield losses were as high as 38 per cent and 43 per cent in the first and second year respectively. The other traits like thousand grain weight and number of grains per spike were reduced up to 15 per cent and 18 per cent respectively.

1.3 Synonyms

Spot blotch has been given different names by different researchers time to time. Name of this fungus was first time given by Pammel et al. (1910) as Helminthosporium sativum but the disease spot blotch, caused by Helminthosporium was reported by Mohy in 1914. The name Helminthosporium sativum by Pammel et al. (1910) was given without taking into account the earlier description of Helminthosporium sorokiniana Sacco in Sorokin, (Trans. Soc. Nat. Univ. Kazan 22: 15, 1890) (Sivanesan 1987). The Ascigerous state (teleomorph) was first observed in the laboratory on natural media in the presence of opposite mating types and described as *Ophiobolus sativus* Ito and Kurib which was later named as *Cochliobolus sativus* (Ito and Kurib) Drechsler Ex. (Dastur 1942). *Cochliobolus drechsler* is characterized by globose ascomata usually with a long cylindrical neck, obclavate-cylindric asci, and helically coiled filiform ascospores. This genus is associated with Bipolaris and *Curvularia anamorphs* and the teleomorphs of *C. sativus* were only reported from Zambia (Raemaekers 1991). Shoemaker (1959) proposed the generic name Bipolaris for the Helminthosporium species with fusoid, straight or curved conidia, germinating by one germ tube from each end (bipolar germination), the spot blotch pathogen was renamed *Bipolaris sorokiniana* (Sacc.) Shoem. This name is presently widely adopted. *Helminthosporium acrothecioides* Lindfors, *Helminthosporium californicum* Mackie and Paxton, and *Drechslera sorokiniana* (Sacc.) Subram. and Jain are considered as its synonyms.

Further, studies have been undertaken on the taxonomy of this group, taking other criteria into account including the type of hilum, the conidial wall and the known or unknown teleomorph, but the terminology proposed by Shoemaker for the above graminicolous Helminthosporium has not been changed (Alcorn 1988).

The international code of botanical nomenclature permits the different states of fungi with pleomorphic life-cycles to be given separate names; if a teleomorph is present, the name automatically refers to that morph even if the anamorph is also present (Hawksworth et al. 1995). The correct name for Helminthosporium leaf blight of wheat is accordingly proposed as: *Cochliobolus sativus* (Ito and Kurib.) Drechsler ex Dastur, anamorph Bipolaris sorokiniana (Sacc.) Shoem. The literature mining reveled that spot blotch pathogen has been given several names time to time (Table 2).

1.4 Symptoms and identification of pathogen on host plant

Fungus infects young leaves by direct penetration into epidermis or through stomata. Initial infection by *Bipolaris* results in small, elongated, dark brown lesions 1 to 2 mm long without chlorotic margin on leaf blades and sheaths. Spots or lesions on leaves are small, chlorotic, and oval shaped with dark center. Reddish-brown centre with yellow margins are formed and ends of the lesions are tapered. Primary infections occur on coleoptiles, primary roots, and sub crown internodes while secondary conidia develop on the above-ground parts of plant. In susceptible genotypes, these lesions extend very quickly in oval to elongated blotches, light brown to dark brown in colour.

Table 2: Names of spot blotch pathogen in chronological order

Synonym	*Reference*
Helminthosporium sorokinianum Sacc.	Ann. Rep. Kansas Agric. Exp. Stn 22(3): 1-32 (1890)
Helminthosporium sativum Pammel, CM King & Blakke	Bulletin of Iowa State College 116: 180 (1910)
Helminthosporium acrothecioides Lindf.	Bot. Tidsskr. 12(2): 212 (1918)
Helminthosporium californicum Mackie & GE Paxton	Phytopathology **13**: 562 (1923)
Ophiobolus sativus S. Ito and Kurib	Trans. Sapporo nat. Hist. Soc. 10: 138 (1929)
Cochliobolus sativus (S. Ito and Kurib.) Drechsler ex Dastur	Indian Journal of Agricultural Research 12: 733 (1942)
Bipolaris sorokiniana (Sacc.) Shoemaker	Can. J. Bot. 37(5): 879 (1959)
Drechslera sorokiniana (Sacc.) Subram & BL Jain	(Subramanian and Jain 1966)
Bipolaris californica (Mackie and GE Paxton) Gornosta- [as 'californicum']	(Bunkina et al. 1978)

They may reach several centimetres in size before coalescing and inducing the death of the leaf. Leaf symptoms are obvious after heading and most frequent on lower leaves. Fruiting structures develop readily under humid conditions and are generally easily observed on old lesions. If spikelet's are affected, it can result in shrivelled grain and black point, a dark staining of the embryo end of the seed. Diseased seedlings develop dark brown areas on the seed and/or sub crown internodes stems below the soil line. Small brown to dark brown spots and discolouration may be observed on head tissues and individual seeds. Although they are similar in colour, spot blotch lesions do not have the typical long, narrow appearance of net blotch lesions. However, spot blotch may be confused with the spot-form of net blotch, which is characterized by small circular to oval, dark brown spots. Severe infections, especially of the flag leaf and second leaf, reduce the amount of leaf area available for photosynthesis resulting in reduced yield and grain filling. For microscopic identification of spot blotch, the slide is made from the infected tissue. The spores are typically boat shaped with multinucleate characteristics. The spores are deep olive or olive-brown, ovate to oblong, erect or slightly curved.

1.5 Disease assessment

Various approaches have been practiced to score spot blotch disease. First approach given by Van der Plank (1963) is based on percentage of diseased leaf area which can be used as a parameter or to calculate the AUDPC (Area Under Disease Progress Curve). It is easy to teach, to manipulate mathematically, and to analyse. In this method, disease severity is assessed 4-5 times at different time intervals by observing the top two leaves and a score on 0 (no disease) to 100 (fully diseased leaves) scale is given. This scoring method is the most universally applicable and particularly useful for comparing epidemic severity across locations and seasons, provided that similar checks are used. On the other hand, it is time consuming. The AUDPC (Per cent day) measures the amount of disease as well as the rate of disease progress. AUDPC value of each genotype is calculated using following formula:

$$\text{AUDPC} = \sum_{i=1}^{i}[\{(Y_i + Y_{(i+1)})/2\} \times (t_{(i+1)} - t_i)]$$

Where, Y = disease severity at time t_i, $(t_{(i+1)} - t_i)$ = Time (days) between two disease scores

The second approach used to screen spot blotch was given by Saari and Prescott (1975). In this approach, plants are evaluated on 0-9 scale. '0' score is given to the plants where no visible symptoms are present while '9' is given to the plant where more than 80 per cent leaf area is under the disease. The recording is based on visual observation. The 0-9 scale is probably very realistic method.

The third method of spot blotch evaluation is the double-digit scale (00-99) developed as a modification of Saari and Prescott's scale (Saari and Prescott 1975) for assessing severity (Eyal et al. 1987). The first digit (D1) indicates disease progress in canopy height from the ground level, while the second digit (D2) refers to severity measured based on diseased leaf area. Both D1 and D2 are scored on a scale of 1 to 9. Since the rate of disease progress in the field can be extremely high in some regions, it is often needed to take repeated scores to properly assess the level of resistance (Dubin et al. 1998; Duveiller et al. 1998). It is recommended to take several individual disease scores per plot at 3 to 7 days intervals over a 3 to 4 weeks period between anthesis and dough stage depending on seeding date. For each score, percent disease severity is estimated based on the following formula: Per cent severity = ((D1/9) x (D2/9) x 100). Lesion size may also be useful in some situations. When adult heads are evaluated; a scale such as "lesion type" can be used quite effectively, particularly under controlled conditions. The advantage of the double digit scale is that the individual components are used to produce an index. This approach, extensively used by some researchers, is quick and easy. However, when genotypes are evaluated under artificial epiphytotic conditions, the double digit scale may not represent correct level of resistance. For example, under artificial inoculation, the inoculum is sprayed over the plant and the upper leaves showed symptoms first than lower leaves. Hence the first digit (D1) of score is always 7-9 even in moderately susceptible genotypes.

More recently, a sensor-based evaluation approach, involving a hand-held green seeker normalized difference vegetation index (NDVI), has been suggested to screen a large number of lines (Kumar et al. 2016). NDVI is a simple graphical indicator that can be used to analyze remote sensing measurements and to assess whether the target contains live green vegetation.

1.6 Variability in the pathogen population

The population of spot blotch pathogen *Bipolaris sorokiniana* found to be variable based on the morphological and cultural properties. In the culture plates, the pathogen ranged from black to white (Maraite et al. 1998; Chand et al. 2003). Hetzler et al. (1991)

reported 15 pathotypes of *B. sorokiniana* collected from different parts of the world. Maraite et al. (1998) further conrmed the evolution of *B. sorokiniana* toward more aggressiveness. Pathological variability has also been reported earlier by Mishra (1981), but it was not reliably correlated with the morphological or any other form of variability. Chand et al. (2003) attempted to group different isolates of *B. sorokiniana* based on their morphological characters and aggressiveness. Several workers across the world, established morphological variability in the population of *B. sorokiniana* (Maraite et al. 1998; Oliveira et al. 1998; Poloni et al. 2008) aiming that it could be useful in monitoring the pathogen population if correlated with pathological variability. However, it is difficult to imply this information practically. Chand et al. (2003) grouped the spot blotch isolates in 5 groups namely, black, brown/dull black, grey with white spots, dull white/greenish black and white based on the morphological characters viz., colony morphology (colour and growth behaviour), radial growth and sporulation on a culture medium. Chand et al. (2003) recorded the frequency of different morphological groups of pathogen at different stage of disease development. They recorded two different frequencies, green to dark green and identify these populations as the main constituents of epidemic. It was noticed that under field conditions only dark green to green population got advantage of faster growth and due to higher sporulation. The researchers observed difference in the pathogenicity among the isolates and found selection for aggressiveness over years (Hetzler et al. 1991; Fetch and Steffenson 1994; Maraite et al. 1998; Poloni et al. 2009). Kumar et al. (2007) studied the pathogenic variability among the spot blotch groups characterized previously by Chand et al. (2002) and it was observed that the black group is the most aggressive. This black group has been further characterized with ITS primers and unique identification number has been obtained for reference. PCR based molecular marker RAPD has been successfully used to identify strains (Guzmán et al. 1999), to characterize races (Malvick and Grau 2001), and to analyse virulence variability related to genetic polymorphism (Chen et al. 1995). Aggarwal et al. (2010) differentiated 40 isolates of *B. sorokiniana* obtained from different geographic regions of India and broadly three major groups were obtained where some isolates showed less than 50 per cent similarity. One of the markers namely URP-2F (52-GTGTGCGATCAGTTGCTGGG-32) amplified three polymorphic fragments which could be used as specific markers for identification of *B. sorokiniana* isolates.

Andres et al. (2001) studied the genetic variability in *Bipolaris* population to determine the amount of intraspecific genetic variability and to study host-pathogen interactions. The resistant, moderately resistant and susceptible plants were inoculated with 10 different isolates collected from Brazil. When the isolates were recovered from the host plants, there was high level of genetic variability among the isolates as tested with RAPD markers. The polymorphism was high and varied, even for the same isolate, when the isolates were recovered from different cultivar hosts.

1.7 Source of variability in the pathogen

According to Burdon and Silk (1997), main source of genetic based variation in plant pathogenic fungi are mutations and recombination. Within a species, gene flow between populations supplements the processes of mutation and recombination as propagules spread from one epidemiological area to another, and from one deme to the next. Gene flow, along with other evolutionary forces, can result in the spread of genes (or DNA sequences), genotypes, and even in the establishment of whole populations in different regions (McDermott and McDonald 1993).

Parasexual recombination and anastomosis have been reported as the cause of variability in the natural populations of this pathogen (Tinline 1962). Chand et al. (2003) suspected this to be due to different combinations of nuclei. In a study of clonal variability of spot blotch pathogen *B. sorokiniana* by Pandey et al. (2008), it was concluded that variability in *Bipolaris* might be generated either through hyphal fusion, nuclear migration, or occurrence of a multinucleate state. Such evidences of microscopic evaluation of nuclei assays also leave little scope to believe that variability could be generated through cytoplasmic factors. Anastomosis is believed to facilitate protoplast fusion and generate multinucleate condition in the cells. However, no direct evidence is available with respect to monitoring of nuclear behaviour in *Bipolaris*. Random Amplified Polymorphic DNA (RAPD) has been successfully used to identify strains and to analyse virulence variability related to genetic polymorphism in pathogenic fungi (Pandey et al. 2005). It is also concluded that the morphological and pathological variability is directly related and morphological feature may be used as marker for a potent disease causing group. The isolates responsible for epidemics and those acting as reservoir for generating new aggressive isolates appeared to be different. Therefore, breeding strategies should take care of both of these populations. In other words, variation-generating isolates need to be constantly observed to face the threat posed by any new potent isolate.

1.8 Possible Impact of Climate Change on the Importance of Spot Blotch

In several parts of the world, the mean winter temperature has risen at least by 0.66°C since 1990 while further warming of approximately 2.13°C by 2050 is expected (Poulton and Rawson 2011). The severity of spot blotch and other foliar disease of wheat may increase in future because a warm and humid climate favourable for the development and spread of the pathogens. Several years back, it is believed that the level of resistance in high-yielding wheat genotypes is still not satisfactory in warmer humid regions of South Asia (Joshi et al. 2007 a, b, c). The Global Wheat Program (GWP) of CIMMYT is giving emphasis on spot blotch resistance breeding by including additional location for spot blotch resistance testing. Consequently, the level of resistance in CIMMYT wheat lines has been increased significantly.

Spot blotch is expanding into the cooler, traditional irrigated rice-wheat production areas which may be due to pathogen adaptation, change in the pathogen population dynamics, reduction in tillage practices, availability of broader host-range, or increased average temperature during crop season. Since, wheat is grown at largest area worldwide, it is likely that climatic and atmospheric changes have an impact on wheat yield and quality (Ortiz et al. 2008; Bender and Weigel 2011). Generally warmer temperatures in winter and throughout the growing season accelerates pathogen life cycles, sporulation and infection efficiency of fungal foliar pathogens. The study on potential future risk through wheat disease was conducted by Juroszek and Tiedmann (2012) and reported that due to shift in warming and other climatic conditions such as altered precipitation, various changes like 1) change in seasonal phenology (e.g. synchronization of pathogen life-cycle stages with their host plants and natural competitors or synergists), 2) change in population dynamics (e.g. over-wintering and survival, changes in the number of generations), and 3) change in geographic distribution (e.g. range expansion or retreat, and increased risk of pathogen invasion) (Melloy et al. 2010; Ghini et al. 2012; Pautasso et al. 2012; West et al. 2012) may occur. Necrotrophic pathogens including *Bipolaris* have always been a major constraint to wheat production worldwide, but unlike rusts, internationally coordinated research efforts are not common for these pathogens. Among other necrotrophic pathogens, severity of spot blotch caused by *Bipolaris* has increased in South Asia with higher average night-time temperatures (Sharma et al. 2007). West et al (2012) concluded based on the epidemic type (epidemic cycles), dissemination method (rain, wind, splash), primary inoculation (air, seed or soil-borne), plant biomass, canopy and weather conditions that polycyclic foliar fungal diseases are predicted to increase in severity.

Until late 1970s, leaf rust caused by *Puccinia triticina* Eriks in Bangladesh was most important disease (Prescott et al. 1986) but at present, *Bipolaris* leaf blight is the most important disease (Joshi et al. 2007 a, b, c). This change may be due to several factors including, replacement of leaf rust susceptible varieties with resistant varieties, less attention was given on development of spot blotch resistant varieties, change in the pathogen population dynamics and adaptation and change in the environmental conditions which favour spot blotch pathogen. The change in environmental conditions may be one of main reasons behind the spread of spot blotch in Bangladesh because the mean temperature of wheat growing season has risen 0.2-0.5 0C since 1951 (http://www.giss.nasa.gov).

Climate change may modify the range and intensity of prevalent wheat diseases in some regions. Therefore, it is possible that economically unimportant pathogens may turn into potential threats in the future (Sharma et al. 2006). Chakraborty et al. (1999) also advocated that climate change may have positive, negative or neutral impact on diseases. Thus, on a global and/or national scale, a particular wheat disease might not increase its future biological and economic importance consistently. However, Chakraborty et al. (2011) reported that the pathogen may adapt to warmer conditions rapidly. The adaptation of pathogens to a changing climate may prove to be one of the most important challenges under future climates (Garrett et al. 2006)

1.9 Biology of *Bipolaris* Reproduction

In nature, *B. sorokiniana* reproduces asexually and the different population can be presumed to be clones. The reproduction in the *Bipolaris* at most places occurs through asexual means except in Zambia (Chand et al. 2003). Although Tinline and Dickson (1958) reported mating types for sexual reproduction, Valjavec-Gratian and Steffenson (1997) reported that the mere presence of two mating types is not enough for sexual reproduction, if other factors were limiting. Wen and Lu (1991) determined the mating type among hundreds of *Bipolaris* isolates collected from different areas and different host plants and observed the prevalence of both 'A' and 'a' mating types. Further study on sex reactions of different isolates specifically from field conditions, revealed that there were no geographic and host limitations on different mating types.

1.10 Survival of Pathogen over season

Several studies have been conducted on the re-occurrence of spot blotch on the wheat crop. Some workers (Shaner 1981; Reis and Forcelini 1994) have found that infected seeds is the main source of inoculum, whereas others (Chinn and Ledingham 1958; Reis and Santos 1987) have reported the survival of free dormant conidia in soil for more than a year in monoculture systems. On the other hand, Ledingham et al. (1960) reported that *B. sorokiniana* survive in the soil not more than 8 months. In India, the monsoon arrives during late June or early July and from here onwards, generally water-logged conditions arise in the rice fields. Water-logging causes anaerobic conditions detrimental to the survival of conidia. The limited number of conidia (<1 per cent) surviving after flooding carry far less potential to cause an epidemic in comparison to high seed infection (12–86 per cent). Further, the conidia isolated from soil in the months of August and beyond are not pathogenic. Therefore, a few viable saprophytic conidia are present and could not be expected to serve as the main source of inoculum in the subsequent wheat crop. Alternatively, it is also possible that grass hosts may harbour the pathogen and act as alternate hosts (Misra 1973; Mishra and Chourasia 1976; Malaker et al. 2008). Duveiller and Gilchrist (1994) suggested that alternate hosts could act as a green bridge for the survival of *B. sorokiniana* across crop seasons. Rice has also been reported as a host species for the spot blotch pathogen in inoculation trials in India (Misra 1973). In a recent study, only three out of 25 plant species associated with rice–wheat cultivation were found to harbour *B. sorokiniana* when investigated in the off-season for wheat (Pandey et al. 2005). It was reported that *B. sorokiniana* isolates from these hosts were pathogenic to wheat under artificial inoculation conditions, their spread to wheat could not be confirmed under natural conditions. Therefore, spread of inoculum from alternate hosts does not appear to be a mechanism for the reappearance of spot blotch in rice-wheat cropping systems after the period of water-logged conditions.

Shaner (1981) reported the survival of dormant conidia of *B. sorokiniana* in seed while Reis and Santos (1987) emphasized that the infected seeds are the only source of inoculum in a new area, even within a season. A small amount of inoculum was sufficient to create an epidemic. In the re-occurrence study of spot blotch infection by Pandey et al. (2005), it was found that the percentage infection of *B. sorokiniana* in the seeds of wheat genotypes was high and the occurrence of spot blotch under non-inoculated conditions in the field and poly-house suggested that infected seed is probably the main source of inoculum for the survival and reappearance in rice–wheat cropping system. Seed produced in the hot spot for spot blotch of wheat, is likely to carry a sufficient inoculum load to cause further spread of disease under favourable conditions.

1.11 Host Range

Bipolaris sorokiniana is a pathogen of broad host range (Kline and Nelson 1963; Misra 1973; Neupane et al. 2010) often infecting a large number of grasses include: ***Triticum aestivum*** (bread wheat), ***Triticum turgidum*** (durum wheat), ***Triticale, Secale cereale*** (rye), T. *vulgare, Zea mays* (Poaceae), Guzmania sp. (Tillandsioideae). *Lacromani anum, P. typhoides, Phleum pratense, Setaria italica, T. durum, T. sativum, T. secale,* ***Hordeum vulgare*** (barley) and wild grasses (Manamgoda et al. 2011). Not only wheat and other plants (Misra 1973) but *Phalaris minor* Retz (Singh et al. 1998b) has also been reported as host species for the spot blotch pathogen. Bakonyi et al. (1998) studied the monocotyledonous hosts of *Bipolaris* by collecting diseased leaf samples from 18 different species (**Table 3**).

Table 3: Monocotyledonous hosts of *Bipolaris sorokiniana*

Sl. no.	*Hosts*	*Sl. no.*	*Hosts*	*Sl. no.*	*Hosts*
1.	*Agropyron pectinatum*	7.	*Bromus inermis*	13.	*Lotiumperenne*
2.	*Agropyron repens*	8.	*Dactytis glomerata*	14.	*Pennisetum villosum*
3.	*Alopecurus pratensis*	9.	*Festuca heterophylla*	15.	*Poapratensis*
4.	*Avena sativa*	10.	*Festuca ovina*	16.	*Secale cereale*
5.	*Beckmannia eruciformis*	11.	*Hordeum murinum*	17.	*Seteriaviridis*
6.	*Bromuserectus*	12.	*Hordeum vulgare*	18.	*Triticumaestivum*

In northeastern China, *B. sorokiniana* found to infect 29 Gramineae and other crop species (Chinkal and Qiben 1982). In the Yellow and Huai River plain region, more than 65 gramineae plants are hosts of *Biopolaris sorokiniana* (Qixiang and Jiabi 1987) while in Guangdong Province of southern China, 17 plant species are reported to be infected by *Biopolaris* (Baichai and Yugan 1983). Among the infected plant species, *Saccharum officinarum, Zizania caduciflora, Paspalum thumbergii, Ischaemum ciliare,* and *Apluda mutica* were the first hosts reported in China.

1.12 Epidemiology

Bipolaris sorokiniana is reported as a weak parasite and colonizes first on older leaves close to the ground or stressed tissues. The amount of inoculums in the field is largely depends on the amount of senescent leaf tissue available for the growth and reproduction of the pathogen. Leaf senescence and spot blotch both progress from lower to upper leaves (Brandle et al. 1987; Raemaekers 1988).

Spot blotch severity and its development depends on various factors e.g., tillage or surface seeding, number of irrigations, soil fertility level, planting density, growth stage of plant, rain during crop cycle (late in season), heat stress owing to late planting, higher temperature and higher humidity etc. (Nema and Joshi 1973; Luz and Bergstrom 1986; Saunders 1988; Reis 1991; Sentelhas et al. 1993; Duveiller 2004; Sharma-Poudyal et al. 2005). Higher relative humidity favours long duration (>12 hours) of leaf wetness. Longer the leaves stay wet, higher the rate of secondary infection because the spores have enough moisture to germinate and penetrate the leaf tissues. High relative humidity arising from high levels of soil residual moisture along with foggy days allows long hours of wetness on leaf blades that can last until late January as studies in the Indo-Gangetic Plains, creates ideal conditions for the establishment and multiplication of wheat pathogen (White and Rodriguez-Aguilar 2001; Lespinay 2004). This situation arises some time even at the end of the monsoon and in the absence of rainfall.

Using Brazilian wheat varieties, Reis (1991) reported that, if wheat leaves remain wet at a mean temperature of 18°C or higher for more than 18 h, foliar blight outbreaks may occur. Moderate to warm temperatures (18°C to 32°C) favours the growth of *B. sorokiniana*. In Asia, Nema and Joshi (1973) and Singh et al. (1998) reported that infection is more rapid and more severe at 28°C than at lower temperatures. In addition to the temperature and humidity, planting time reported to plays a significant role for disease severity. For epidemiological studies, significantly higher AUDPC values are obtained if planting is done during the last week of November to last week of December (Duveiller et al. 2005a). The higher values of AUDPC under late-sown conditions are most likely caused by heat stress, which probably weaken the plant defence system favouring disease development (Nema and Joshi 1973; Duveiller and Sharma 2009). Delayed seeding for wheat, grown after rice in eastern India and Nepal also results in higher losses of grain yield and total kernel weight due to heat stress as well as foliar blight (Hobbs and Giri 1997; Singh et al. 1998b; Duveiller et al. 2005b).

1.13 Mode of infection by *Bipolaris*

As soon as the conidia come in contact of leaf surface, the adhesion process starts. Conidial adhesion is mediated by weak hydrophobic forces and occurred only on hydrophobic surfaces. The next stage of adhesion is firm, surface unspecific and accompanied with conidial germination. After the the germ tube adhesion, the fungus produces an extracellular matrix (ECM). Pathogenesis includes germination of conidia on the leaf surface and formation of an appressorium at the tip of the germ tube that supports direct penetration through the host cuticle. The apparently biotrophic growth phase is primarily confined to a single epidermal cell invaded by infection hyphae, whereas the necrotrophic growth phase starts upon invasion of the mesophyll tissue followed by host cell death, which appears to be a consequence of toxin secretion fungal penetration into the epidermal layer failed mostly through a cell wall-associated defence mechanism. In cases where the fungus successfully overcame epidermal defense, its spread within the mesophyll tissue (necrotrophic phase) is restricted in the more resistant genotypes. Epidermal cell wall-associated defense, spreading as well as the extent of electrolyte leakage of infected tissue, correlated well with field resistance (Ibeagha et al. 2005).

Nutrients which leach out from the healthy wheat leaf tissue are absorbed by the conidia and utilized for further penetration (Yadav 1981). In the tropics, inocula are generally found in the soil round the year (Duczek 1981). Along the germination of the diseased seeds, the perennating organs of the causal organism become active. This is the starting point of the disease. Conidia germinate completely in four hours followed by appressoria formation at the juncture of epidermal cell wall in eight hours. Hyphae from initially infected cells enter adjacent cells in 24 hours, which results in the granularisation of the host cytoplasm (Bisen and Channy 1983). The fungus is then transmitted to the plumules and coleoptiles tips with an efficiency reaching up to 87 per cent (Reis and Forcelini 1994). Maximum development of symptoms appears when the leaves remain wet for more than 18 hours with a mean temperature greater than 18 ^{0}C (Couture and Sutton 1978). Disease development is favoured by periods of overcast, high humidity with fog or rain and high temperatures (18 to 28 C). Leaves of seedlings and tillering wheat plants can become severely diseased during warm wet days (Duveiller et al. 2007). However, farmers may not notice leaf spots until flag leaf emergence or early heading. Under favourable conditions, hyphae produce conidiophores, which emerge out through stomata of the host tissue. The emerging conidiophores produce a succession of conidia, which are transmitted by rain splashes and wind, thus building up polycyclic epidemics. Conidia on germination produce germ tube, which is surrounded by thick mucilaginous substrata. This mucilaginous substratum enables the germinating conidia to remain adhered to the host surface. The germ tube then swells to produce appresorium from which infection hyphae are developed. The infection hyphae then enter the host tissue either through stomata or by rupturing through epidermis. Immediately after the entrance in the host tissue, the infection hypha divides rapidly and ramifies along the intercellular spaces of the mesophyll tissue but restricted Mode of infection and disease cycle of Bipolaris sorokiniana causing spot blotch disease in wheat is explained in Figure 1.

1.14 Disease Cycle

Biotrophic development of *Bipolaris* without symptoms occasionally results in enhanced senescence of all leaves (Dehne and Oerke 1985). Although, *Bipolaris* is seed borne pathogen but the primary infection can be through infected crop residues, collateral hosts and free dormant conidia in soil in addition to the seed (Reis 1991; Chand et al. 2002). Survival of *Bipolaris* in seeds is probably the major source of primary inoculum (Mehta 1993). According to Shaner (1981), it is the main source of inoculums of leaf blight pathogens. If the seeds are infected when they sow, it provides inoculum for the newly grown crop (Reis 1991). The fungus also overwinters in the soil and crop residues and infects young seedlings in the spring. The trends of survival duration in wheat crop residues and infected soil is more or less similar and it was found that the population of *Bipolaris* increased initially for 2 months in both soils and crop residues and declines thereafter (Malaker et al. 2008). However, conidia may survive in soil for more than 2 years (Ledingham et al. 1960). The sexual stage is not important in the disease cycle.

Conidia are the main dispersal and survival propagules (Reis and Wünsche 1984). It has been observed that the parasitic conidia on wheat straw can form conidial aggregates. These clumps of conidia are morphologically similar to sclerotic (Chand et al. 2002). Secondary infection may be caused by airborne conidia (Duveiller et al. 2005a) and this increases the severity of the disease. Nutrients available in the conidia of *Bipolaris* are sufficient for germination and formation of aspersoria; however, conidia need an exogenous supply of nutrients to enable successful penetration (Yadav 1981).

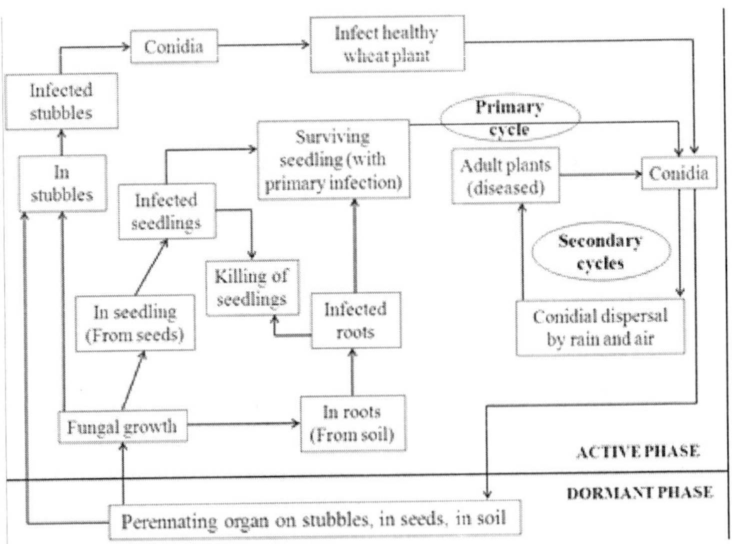

Figure 1: Mode of infection and disease cycle of* Bipolaris sorokiniana *causing spot blotch disease in wheat [Adapted from Acharya et al. (2011)]

1.15 Toxins produced by *Bipolaris*

Phytotoxins are low molecular weight secondary metabolites (sesquiterpene) induces both chlorosis and necrosis (Carlson et al. 1991) causing plant cells death at very low concentrations, of nanomolar (nM) to micro molar (Berestetskiy 2008). There are several reports on the presence of phytotoxic substances in cultural filtrates of B. sorokiniana (Ludwig 1957; Gayad 1961). The phytotoxins produced by spot blotch pathogen are prehelninthosporol, helminthosporol, helminthosporic acid and sorokinianin (Carlson et al. 1991; Liljeroth et al. 1994). More than 20 compounds related to helminthosporol have been isolated from different species of the genus Bipolaris (Chavesa et al. 2006). Prehelninthosporol (C15H22O2) is a hydrophpbic sesquiterpene with restricted thermal stability and low water solubility; it is detected from conidia and hyphae (Carlson et al. 1991). Prehelminthosporol considered to be the most abundant and active compound of *B. sorokiniana* and may be produced up to 2.1 μg/mg of dried fungal tissue (Carlson et al. 1991). The mechanisms of action have been determined only for limited number of Phytotoxins. Phytotoxins leads to suppression of cell division, physiological disorders (e.g., activation or inhibition of respiration), biochemical changes (disruption of compound synthesis or transportation), and damage of cell membranes (resulting in ion efflux). The main molecular targets for phytotoxins are components of cell membranes (structural or transport proteins and lipids) and enzymes. The affected plants are characterized by the appearance of necrotic spots on the surface of the leaves. Helminthosporol (produced by Bipolaris sorokiniana) increases considerably the membrane permeability to protons and substrate anions in mitochondria, microsomes and chloroplasts. It also inhibits oxidative phosphorylation in mitochondria, photophosphorylation in protoplasts and the operation of the proton pump in the plasma membrane (Berestetskiy 2008). Sorokinianin is a novel fungal metabolite (sesquiterpene) composed of a C3 unit plus the same carbon skeleton as Prehelninthosporol. Victoxinine has been reported from the filtrates of Bipolaris sorokiniana (Pena-Rodriguez and Chilton 1989). Helminthosporol is a plant growth regulator that promotes shoot growth in rice seedlings but inhibits the growth of wheat seedlings (Tamura et al. 1963; Tamura et al. 1965). Helminthosporol and the associated metabolite helninthosporol are most probably artefacts arising from the precursors Prehelninthosporol (octahydro-5-isopropyl-8-methyl-9-methylene-4, 8-methano-1H-2-benzopyran-3-ol) and Prehelninthosporol, respectively (Mayo et al. 1965).

1.16 Mechanism of phytotoxins resistance

The appearance of symptoms heralding the onset of phytotoxic (PT) induced diseases precedes fungal spore germination, penetration by the infectious hyphae of the host tissue, and the development of the mycelium in it. Depending on the fungal species, PT synthesis may occur at any of the stages of pathogenesis, listed above. The majority of PTs act locally; however, highly mobile phytotoxic substances are also known (examples include fusaric acid and the PT of the fungus *Ascochytac hrysanthemi*). Various elements of passive immunity of plants (e.g., morphological features of the plant surface, phytoncides, and papilla formation) prevent pathogens from entering the tissues and developing therein, and thus rule out the possibility that the toxins contact the plant cells. Immune-labeling made it possible to detect prehelminthosporol, in the Voronin bodies of quiescent conidia and the extracellular matrix of the growing hyphae of the fungus. This suggests that the pathotoxin enters the tissues of barley leaf following disruption of its cuticle by the fungal enzymes. Although the toxin of *B. sorokiniana* is nonspecific and highly active against plants of diverse families, only cereals are the major hosts of the producer pathogen. Certain soil fungi are capable of releasing phytotoxic substances directly into the soil; the toxins affect the plants by acting on their root system. The higher the activity of glutathione S-transferase, the greater is the resistance of these cells to the toxin (Coskun and Zihnioglu 2002). Plants resistant to *Alternaria brassicae* (e.g., *Sinapis alba*) are capable of degrading the toxin it produces (destruxin B) (the degradation process involves by hydroxylation followed by glycosylation). In addition to inactivating the toxin, this causes phytoalexin induction. 2, 5-Anhydro-D-glucitol (AhG; a structural analogue of fructose), produced by 18883 *Fusariusolani* strain NRRL 18883, exhibits phytotoxic activity in plants when phosphorylated by plant enzymes to AhG-1,6-*bis*P. AhG-1,6-*bis*P (the biologically activated form of AhG) completely inhibits aldolase, which causes the susceptible cells around the infected site to die. Thus, the plant insensitivity to PTs may be associated with their inactivation by plant enzymes or the absence of biological activators. Plant cells are insensitive to PTs in the absence of appropriate molecular targets. If the targets are available, plant susceptibility may strongly depend on their structural and conformational integrity; indeed, it is possible to block the effects of a PT by changing the structure and/or conformation of its targets (and thereby rescue the plant, unless such treatment is deleterious). Such approaches would affect the operation of plant genes (expression, introduction of mutations, etc.) associated with the synthesis of the PT target. For example, maize cultivars devoid of the gene URF13 are insensitive to T-toxin.

1.17 Morphological markers and traits associated with spot blotch disease resistance

1.17.1 Plant height and days to maturity

Resistance of spot blotch is considered to be associated with late growth stage of the plant irrespective of days to maturity, height and other observable characters (Duveiller and Gilchrist 1994). Duveiller and Gilchrist (1994) reported that taller genotype may escape disease due to greater interned length and slower movement of spore to the upper leaves. The spot blotch incidence was reported to be independent of plant height and days to maturity by Joshi et al. (2002) after investigating 1407 spring wheat (*T. aestivum*) lines of Indian and CIMMYT origin for plant height, days to maturity and resistance to spot blotch. Rosyara et al. (2009) studied three different spring wheat populations and also observed week correlation ranges from -0.29 to -0.43 (significant or non-significant) between spot blotch incidence and plant height as well as with days to heading. It was suggests that a genetic association between resistance and taller plant height or later maturity is not always true. The low level of disease in early maturing lines may be due to a kind of escape mechanism rather than genetic resistance. Similarly, the less disease severity on taller plants is due to slower secondary infection. Therefore, it is possible to find resistant plants that are dwarf and early. Further, in genetic studies or in screening of exotic sources of resistance, it is important to ensure that apparent resistance is not a result of increased plant height or delayed flowering. The doubled haploid population developed using CIMMYT's derived synthetic wheat lines SYN1 [Mayoor//Tksn1081/*Ae. squarrosa* (222)] resistant to spot blotch and Ocoroni (highly susceptible elite line) was investigated for correlation between spot blotch AUDC, plant height and days to flowering. Highly significant but negative correlation was found between spot blotch susceptibility and plant height or days to flowering ranging from -0.30 to -0.53 (Pawan Singh, CIMMYT, personal communication).

1.17.2 Leaf angle

The other trait called leaf angle seems to be associated with spot blotch resistance. Joshi and Chand (2002) studied more than 1400 germplasm lines and concluded that the lines with erect and semi-erect leaves displayed lower spot blotch severity than those having drooping and semi-drooping leaves. The low spot blotch score of erect leaf genotypes and a positive phenotypic correlation between leaf angle and AUDPC (+0.58) suggested that erect leaves contributed positively to reducing the severity of spot blotch disease. However, the presence of both high and low AUDPC types in erect and drooping lines indicated the absence of complete genetic linkage. Other indirect effects may also have contributed. The low mean AUDPC scores of erect leaf progenies might partly be due to the fact that erect leaves retain less free water that is essential for germination of pathogen spores.

1.17.3 Leaf tip necrosis

Joshi et al. (2004) reported leaf tip necrosis (Ltn) as morphological marker to select spot blotch resistant plants. The plants possess Ltn showed resistance to spot blotch disease. LTN of wheat, which is known to be due to the presence of gene Ltn, is reported to be linked with genes Lr34 and Yr18 involved in slow rusting resistance to leaf [*Puccinia recondita* Roberge ex Desmaz. f. sp. *tritici* Eriks & E. Henn.) D.M. Henderson] and stripe rusts (*Puccinia striiformis* Westend.), respectively (Singh 1992a; Singh 1992b; McIntosh 2007).

1.17.4 Stay green

Stay green is also known to display good association with resistance to stem rot (Rosenow 1984) suggesting that stay green leaves remain photosynthetically active even under biotic stress conditions. By maintaining green coloration, stay green trait may resist the growth and development of other diseases especially those caused by facultative pathogens such as the spot blotch pathogen (Mercado et al. 2003). Joshi et al. (2007a) observed that there is positive association of the stay green trait and spot blotch resistance that may be exploited for developing spot blotch resistant wheat.

1.18 Spot blotch resistance and heat tolerance

Temperature is one of the factors associated with increased infection and disease development caused by *B. sorokiniana* (Duveiller et al. 2005a). The maintenance of healthy tissues by plants enables them to maintain a lower canopy temperature which, in turn, helps against the pathogen which requires relatively high temperatures for faster growth and development. Spot blotch resistance and heat tolerance stresses are supposed to be associated complicating development of tolerant genotypes. To assess the potential application of canopy temperature depression (CTD) as an integrative trait for screening spot blotch resistance and heat stress tolerance, Rosyara et al. (2009) used ten genetically diverse genotypes with different level of resistance to spot blotch. The lines were grown both heat stressed (late sowing) and non-stressed (timely sowing) field conditions. They observed strong negative correlation between AUDPC per day and area under canopy temperature depression progress curve (AUCTDPC, r = - 0.72). It indicates that foliar blight susceptibility has important role in decreasing AUCTDPC. Heat stress alone is also expected to reduce AUCTDPC values in late-sown wheat, thereby reducing the correlation between the two parameters. Obviously, heat stress may have a confounding effect with the symptoms of foliar blight (Duveiller and Gilchrist 1994) or by increasing the temperature of plants without symptoms of senescence. Both components are expected to have adverse effect on photosynthesis. Other researchers (Pederson and Nutter 1984; Eyal and Blum 1989; Rosyara et al. 2010) also reported that CTD is reduced due to disease for different host-pathogen systems.

1.19 Source of spot blotch disease resistance

Several sources of spot blotch resistance have been identified in wheat (Duveiller and Gilchrist 1994; Velazquez-Cruz 1994; Dubin and Rajaram 1996), but it is generally accepted that resistance of high yielding genotypes is not yet satisfactory. Van Ginkel and Rajaram

(1998) reported that there are three categories of resistance sources to spot blotch viz., Latin America, Chinese and wild relatives of wheat or alien species. The Chirya lines displaying good level of resistance were developed using *Thinopyrum curvifolium* by Mujeeb-Kazi et al. (1996). The Latin American sources are mainly derived from Brazil. Resistant Brazilian commercial varieties are BH1146 and CNTl (Mehta 1985). The Chinese sources of resistance used at CIMMYT include Shanghai #4, Suzhoe 8, and Yangmai 6 (van Ginkel and Rajaram 1998). Some other Chinese wheat genotype like SW895422, Chirya 1, Chirya 3, Chirya 7, NL781, and NL785 also showed significant resistance levels to spot blotch (Kohli et al. 1991). Wheat genotypes Yangmai 6, M3 (W7976), Shanghai 4, and Chirya 7 showed higher levels of resistance (Ibeagha et al. 2005). Using alien sources like *Thinopyrum curvifoIium* (Mujeeb-Kazi et al. 1996a; Villareal et al. 1996) lines such as Mayoor and the Chirya series were developed at CIMMYT. Mayoor showed high degree of resistance both under field and polyhouse conditions (Khan and Chowdhury 2011). Soma clones developed by Arun et al. (2003) from two spring wheat verities HUW 206 and HUW 234 by using immature embryos also showed a significant resistant against spot blotch. Some species like *Ae. triuncialis, Ae. speltoides, Ae. triaristata, Triticum dicoccoides, Ae. cylindrica, and T. boeoticum* also showed a significant level of resistance for spot blotch (Dhaliwal et al. 1993; Singh and Dhaliwal 1993). An early Mexican variety, Cocoraque, was found to express field resistance for spot blotch (Mehta 1985). Kea and Opata showed adaptation to spot blotch affected regions. *Aegilops squarrosa also* show impressive resistance to spot blotch in Mexico. *Aegilops squarrosa* crosses have shown impressive resistance to spot blotch in Mexico. Chirya 3, BW 14999, Ning 8319, Shanghai 4, Shanghai 158, Longmai, Mayoor, Suzhoe 1–58, Arthur/BICRE/1MR 63, Suzhoe 128-OY, IBWSN 81, Chuanmai 18, Ning 8201, Chirya 7, HD 2819, HW 2093, HP 1729, Longmai 10370 show a significant level of resistant to spot blotch (Khan and Chowdhury 2011). A high frequency of resistant specimens was noted in species *T. timopheevii, T. araraticum, T. boeoticum, T. Persicum*, and *T. Urartu* widespread in Transcaucasia and adjacent countries, as well as in *T. sphaerococcum*, the native land of Indian Peninsula (Smurova and Mikhailova 2007).

T. dicoccoides, T. dicoccum, and *T. Turgidum* highly resistant specimens were found (Smurova and Mikhailova 2007). Very recently, a new wheat variety called Francolin (Also known as BARI Gom 27, previously BAW 1120) was released in 2012. This new variety not only resistant to most deadly fungus UG99 but showed good level of resistance to spot blotch as well (Dr. Ravi Singh, CIMMYT, personal communication). Sharma et al. (2007) reported that the genotype Altar-84/Ae. Sq. (224)//Yaco had lowest disease severity and highest 1000-kernel weight (TKW) and grain yield. It was also the most stable for spot blotch resistance, TKW and grain yield as tested in Nepal. Mahto et al. (2011) developed six advanced breeding lines viz., SW89-5422, BL2127 (Danial88/HLB30//NL297), BL3033, Filin/Irena/5/Cndo/R143 //Ente/Mexi-2/3/Ae.squa (TAUS)/4Weaver, GAN/*Ae.squarrosa* (236)//DOY1/ *Ae.squarrosa* (447)/3/MAIZ /4/Inqalab91, Mayoor//TK SN1081/*Ae.squarrosa*(222)/3/FCT, resistant to spot blotch, *Stagonospora nodorum* blotch (SNB) and tan spot race 1. The resistant wheat lines identified in his study represent potentially useful and untapped sources of resistance to multiple leaf spot diseases.

Joshi et al. (2007c) tested 729 lines for five years at eight locations in lowlands of south Asia. Three lines viz., EGPYT67 (Kauz//Kauz/Star/3/Prinia/4/Milan/Kauz), EGPYT84 (Mrng/Buc//Blo/Pvn/3/Pjb 81) and EGPYT69 (Chirya3/Pastor) were found the best for spot blotch resistance as well as for yield, days to maturity, and thousand kernel weight (TKW).

1.20 Genetics of resistance

One of the major constraints in breeding progress to enhance resistance against spot blotch is lack of detailed information about the inheritance of resistance for this disease. Inheritance studies on resistance to spot blotch are limited and nature of inheritance is still debatable. In general, the degree of resistance in modern cultivars is still insufficient (Hetzler et al. 1991; Mujeeb-Kazi et al. 1996a; van Ginkel and Rajaram 1998; Mujeeb-Kazi et al. 2007). Progress was made by transferring resistance genes from *Thinopyrum curvifolium*, *Elymus curvifolius*, and *Triticum tauschii* to wheat (Villareal et al. 1996; Mujeeb-Kazi et al. 1996b). Further, allelic relationship of the resistant genes is also not clearly known however, some preliminary information on allelic relationship using various source of resistance was provided by Kumar et al. (2007b) and Bhushan et al. (2002). Both qualitative and quantitative inheritance of spot blotch resistance are suggested. Dominant and recessive genes are reported to control inheritance of resistance and in certain cases, epistasis also has been found (Steffenson et al. 1996; Mikhailova et al. 2004; Ragiba and Prabhu 2009). Adlakha et al. (1984) reported that spot blotch resistance in the seedling stage was conditioned by one or two dominant genes while it is under the control of polygenes at adult plant stage. Srivastava et al. (1971) indicated that seedling resistance to spot blotch was controlled by two dominant epistatic genes. Similarly, (Sharma et al. 1997b) found that three dominant epistatic genes were involved in the genetic control of spot blotch. In a field study in Mexico, Velazquez-Cruz (1994) reported that spot blotch resistance was partially dominant, with two to three genes conditioning adult plant resistance. In contrast, others have reported that one to three complementary recessive genes confer resistance to wheat foliar blight (Singh et al. 1998a; Singh et al. 2000; Ragiba et al. 2004). Similarly, Bhushan et al. (2002) reported that two recessive genes with additive effect controlled leaf blight resistance in Punjab, India. Joshi et al. (2004b) reported that three additive genes controlled spot blotch resistance in the eastern Gangetic Plains of India.(Joshi et al. (2004a) also reported that leaf tip necrosis (Ltn), which is under the control of a major gene, is associated with moderate resistance to spot blotch and can be used as a morphological marker to facilitate selection for resistance. The heritability estimates for the three crosses were moderate with a range of the heritability estimates for the three crosses were moderate with a range of 0.77 to 0.84. A few other reports (Sharma et al. 1997b; Singh et al. 1998a; Sharma and Bhatta 1999; Singh et al. 2000; Joshi et al. 2004b; Sharma et al. 2006; Kumar et al. 2007b) also suggested moderate estimates of heritability with respect to spot blotch resistance. Joshi et al. (2004b) suggested that by creating effective artificial epiphytotic and recording observations taking care of the growth stage, one could reduce the effect of environment and thereby obtain fairly high estimates of heritability. Sharma et al. (1997b) also reported that spot blotch resistance in wheat was quantitatively inherited, with intermediate to high estimates of heritability. In other study (Neupane et al. 2007) also reported that the resistance of spot blotch is governed by a single dominant gene. Kumar *et al.* (2007b) reported that around two and three genes were responsible for spot blotch resistance. Kumar et al. (2009) and Kumar *et al.* (2010) identified seven QTls for spot blotch resistance using three different source of resistance. Since genes are only a few and heritability is fairly high, effective selection could be applied in the early segregating generations (Joshi et al. 2004b). For further gains, strong selection pressure could be applied in advance generations when high homozygosity has been achieved (Singh et al. 1995; van Ginkel and Rajaram 1998; Joshi et al. 2007b; Sharma et al. 2008).

1.21 Markers and QTL for blotch resistance

The resistance to spot blotch is controlled not only by a single gene but with more than one gene or QTLs depending on the source of resistance. Further, one source of resistance may have different genes or QTLs combination than other (Kumar et al. 2007b). Limited QTL mapping studies on spot blotch resistance has been conducted in wheat by researchers. Sharma et al (2007) identified three microsatellite markers viz., *Xgwm067*, *Xgwm570* and *Xgwm469* on 5BL, 6AL and 6DS respectively associated with spot blotch resistance using G162 (CIMMYT CID 93926) as resistant parent employing bulked segregation analysis approach. Kumar et al. (2005) identified two microsatellite markers *Xgwm437* and *Xgwm544* on chromosome 7D and 5B respectively using single marker regression analysis. Kumar et al (2009, 2010) used three different RIL mapping populations (Yangmai 6 × Sonalika, Chirya 3 × Sonalika and Ning 8201 × Sonalika) and identified a total of 8 QTLs on different chromosomes.

Recently Lillemo et al. (2013) detected QTLs for spot blotch resistance on chromosome 5B, 7A and 7D. The QTL on chromosome 7D was detected as a major QTL and was of more importance because this QTL is detected in the same chromosome region where the *Lr34* gene is present, this QTL was designated as gene *Sb1* (flanking markers Xgwm1220 – Xgwm295 on chromosome arm 7DS). This indicate that the Lr34 gene complex not only associated with Pm34/Yr18/LTN gene complex (Lagudah et al. 2009) but associated with spot blotch resistance as well. The other minor QTL was mapped on chromosome

1BL corresponding to the *Lr46* between the markers *hbe248* and *ncw1-V* in wheat (Lillemo et al. 2013). This chromosome region on 1BL is reported to be associated with Lr46/Yr29/Pm39 gene complex (Rosewarne et al. 2006).

Two other genes for spot blotch has also been Identified, second gene named as Sb2 (QSb.bhu-5B on chromosome arm 5BL) by Kumar et al. (2015). The third gene Sb3 was identified through bulked segregant analysis (BSA) and was fine-mapped on chromosome arm 3BS in the vicinity of two other QTLs (QSb.bhu3B and QSb.cim3B; Lu et al. 2016).

In addition to the SSR markers, several RAPD markers are reported to be linked with spot blotch résistance (Patil and Hanchinal 2011, Sharma et al. 2007a). Kumar et al. (2004) identified two RAPD markers OPT02 and OPD06 associated with spot blotch resistance where marker OPT02 was found around 10.2cm away from the resistance gene in coupling phase and OPD06 was located 7.7 cm away from the gene of susceptibility in repulsion phase. Recently, Zhu et al. (2014) reported molecular markers for spot blotch resistance on chromosome 1B, 3B and 5A in a synthetic-derived wheat line of CIMMYT.

1.22 Genes for spot blotch resistance

Janni et al. (2008) reported that expression of Bean Polygalacturonase-inhibiting proteins (PGIP) in transgenic wheat has strong defense against the fungal pathogen Bipolaris. Since Bipolaris produces a cell wall glycoprotein, endopolygalacturonases (endo-PGs) which is inhibited by both the bean and the wheat PGIP (Clay et al. 1997; Kemp et al. 2003). PGIP belongs to the subclass of proteins containing leucine-rich repeats (LRRs) of the extracytoplasmic type (Jones and Jones 1997). Wide range of defence responses are elicited due to interactions between pathogens PGs and host PGIP which favours accumulation of oligogalacturonides (Cervone et al. 1997; Ridley et al. 2001). A novel pathogen induced ethylene responsive factor (ERF) gene called TaPIEPI (transcription activator belongs to B-3c subgroup of the ERF family) exhibit significantly increased resistance to Bipolaris sorokiniana (Dong et al. 2010). TaPIEPI is the first wheat ERF protein that is involved in modulating wheat defence response to Bipolaris sorokiniana. TaPIEPI dramatically increases the transcript level of pathogen related (PR) genes (chitinase, B-1, 3-glucanase and PR10) which defend against Bipolaris sorokiniana. Proteins from Chitinase and B-1, 3- glucanase genes inhibit the hyphal growth of Bipolaris sorokiniana.

1.23 Breeding for spot blotch resistance

Cultivation of disease resistant varieties is an economical and effective option for controlling any plant disease. However, conventional breeding of wheat for selection of genotypes resistant for spot blotch has made very limited progress in the past (Sharma et al. 2007b). A few studies from South Asia have reported only low to moderate success in breeding for spot blotch and foliar blight resistance (Alam et al. 1994; Sharma et al. 1997a; Sharma et al. 2004; Siddique et al. 2006; Joshi et al. 2007c). Breeding for resistance to spot blotch has been intensified through International Collaborative Projects (Israel, Germany, CIMMYT and Brazil). Selection for spot blotch resistance in segregating populations started in the early 1980s.

Efforts are being made to develop high yielding wheat genotypes with resistant to spot blotch using breeding skills and pathology resources for the non-traditional warm wheat growing environment. Breeding for resistant in past 20 years has produced limited progress and newly developed wheat cultivars suffers considerable yield loss under spot blotch epidemics in the regions where spot blotch is more frequent. The main strategy to develop disease resistant cultivar is use of highly resistant source with integrated approach. In this strategy, priority is given to the selection of best resistant source. The breeding process starts with the choice of parents to be crossed. Specific crosses are made between high yielding germplasm, adapted varieties, and sources of spot blotch resistance. The first crosses to incorporate spot blotch (caused by *Bipolaris sorokiniana*) resistance into CIMMYT wheats were made about 20 years ago. These crosses involved moderately resistant cultivars, such as BH1146 from Brazil. However, the level of resistance in progenies was inadequate when tests were carried out at Poza Rica, Mexico, CIMMYT's ME5 testing site. In the mid-1980s, wheat genotypes carrying resistance to scab and obtained from the Yangtze River Valley of China, showed varying levels of spot blotch resistance when tested at Poza Rica. These Chinese lines included Suzhoe 1 to 10, Wuhan 1 to 3, Shanghai 1 to 8 and certain Ningmai and Yangmai lines. About the same time, the wide crossing program at CIMMYT produced resistant lines, which contain *Thinopyrum curvifolium* in their pedigree (Villareal et al. 1996). Some of these lines and their derivatives are showing good resistance and appear to be promising in Bangladesh, low-land Bolivia and Nepal and resistance in these wheat's, such as Sabuf, Chirya 1 and Cucap, appear to be controlled by two to three genes (Velazquez-Cruz 1994) whereas Longmai 10 and Yangmai 6 may carry polygenic resistance with high narrow sense heritabilities (Sharma et al. 1997b; Sharma et al. 2006). A few synthetic wheats developed at CIMMYT also carry resistance derived from the *Aegilops tauschii* accessions. A key problem with selection for spot blotch resistance is the negative correlation of disease severity with heading date and plant height (Duveiller and Gilchrist 1994). Therefore, care must be taken if short types with early maturity are required. Identification of some highly resistant lines from such crosses indicate that resistance is additive. International Wheat and Maize Improvement Center (CIMMYT) and National Agricultural Research Centre (NARC), Nepal have been making some effort to incorporate and exploit alien resistance genes in a wheat background using diverse sources of genetic variation in order to pyramid genes (Mujeeb-Kazi 2003).

1.24 Breeding strategy

The breeding strategy for spot blotch needs to rely on the infusion of resistance from alien species such as *Triticum tauschii* (*Aegilops squarrosa*) and *Thinopyrum curvifolium*. Selection for spot blotch resistance must be combined with heat tolerance. There is a need for testing at key sites (at least 5-10) in order to identify suitable parental material and to exchange advanced and segregating materials with good agronomic type and high resistance levels. Tan spot and leaf rust resistance need to be combined; however, a better understanding of the relationship between tan spot development and zero tillage is needed to make breeding for this resistance more efficient. To develop wheat populations with the desirable characters, the use of backcrosses and multiple crosses will be most advantageous, as well as understanding the correlation between greenhouse and field testing. Multi-location testing, shuttle breeding has been successful at CIMMYT and at national centers. It should continue to be emphasized, along with the exchange of germplasm and information on a regional and global scale.

1.25 Marker assisted breeding for spot blotch resistance

In recent past, several molecular markers linked with spot blotch resistance genes/QTLs are made available. Those markers are being used in marker assisted breeding for to combine alleles from different resistance sources. Recently, Indian Council of Agricultural Research (ICAR) India notified Banaras Hindu University as one centre where MAS is being done to develop spot blotch resistant lines.

In one of our breeding program, we made 13 resistant × resistant crosses in bread wheat to combine alleles from different resistant sources. After crossing, advancing the generations and using linked molecular markers, 5-6 advanced lines per cross were selected which carry the alleles combined from both the parents and showed near immunity to spot blotch disease as tested in the field condition under artificial epiphytotic conditions. Those lines may be used effectively in breeding program

1.26 Disease management

Keeping in view the losses caused by B. *sorokiniana* and its expansion toward cooler traditional wheat growing areas, integrated management strategies are needed for spot blotch control. Crop rotation favours beneficial soil organisms as well as promotes better plant nutrition. According to Reis et al. (1998) fungicide treatment and crop rotation with non-host crops can control spot blotch. Late sowing is not recommended, in order to avoid coincidence between the flowering-milk stage and the hot and humid periods during April. Scientist suggests that rotating crops with suitable non-host crops such as rape, sorghum and soybean might reduce the carry over population of the pathogen to very low level and selection of suitable non-host crop requires further research.

Because, seed is one of the important sources of primary infection, seed treatment with suitable fungicides such as Captan, Mancozeb, Maneb, Thiram, Pentachloronitrobenzene (PCNB) or Carboxinguazatine plus are useful in protecting germinating seeds and (Hait and Sinha 1986; Mehta 1993). Sharma-Poudyal et al. (2005) has reported that seed treatment with Vitavax 200B and Carbendazine improves early plant establishment in heavy soil predominating areas where wheat is cultivated after rice.

Mixtures of wheat cultivars with different levels of resistance to spot blotch are a further option (Sharma and Dubin 1996). Avoidance of the economically eligible zero tillage and stubble retaining is an additional measure that supports the control of crown rot and common root rot disease by farmers (Bailey and Duczek 1996; Wildermuth et al. 1997). These observations underline the value of biological strategies in crop protection systems. These strategies must be combined with appropriate plant production measures.

1.27 Biological control

Biological control of this disease (Mandal et al. 1999) includes the use of potential antagonist of several soil and seed borne plant pathogens (Vannacci and Harman 1987; Walther and Gindrat 1988; Pierto 1990) like saprophytic ascomycete and *Chaetomium globosum* Kunze. Recent studies have emphasized the role of *C. globosum* in controlling spot blotch of wheat caused by *Bipolaris* vitro and in vivo conditions (Biswas et al. 2000; Aggarwall et al. 2004). Although, used of chemicals in controlling spot blotch is not environment friendly, the use of the fungicide Triazole group (e.g., Tebuconazole and Propinazole) and Opus (Epoxiconazole) reported to bring disease severity below 10 per cent suggesting its value in controlling spot blotch in the absence of highly spot blotch resistant wheat cultivars (Duveiller et al. 2005b; Sharma-Poudyal et al. 2005).

Kumar et al. (2002) observed that *Piriformospora indica*, a plant growth-promoting root endophytic basidiomycete reduces the number and size of leaf lesions significantly caused by *Bipolaris sorokiniana*. Infestation of wheat roots by this fungus (*Piriformospora indica*) results in a considerable increase in growth and yield relative to non-infested controls. In Canada, where common root rot prevails, rotation, including two or more years of flax (*Linum usitatissimum*) as a break crop, reduced the amount of viable inoculum of *Bipolaris sorokiniana* in the soil (Malaker et al. 2008).

The other promising strategy to manage spot blotch is induced resistance. In broad terms, induced resistance is the control of disease by a prior activation of the plant's own defence system which may be activated by necrotizing pathogens as well as by chemicals mimicking factors of the natural defense systems, such as salicylic acid (Sticher et al. 1997; Bochow et al. 2001). The efficacy of the chemicals in inducing resistance is rather unpredictable and resistance strategies by chemical inducers are inapplicable for farmers so far.

It has been reported that, pretreatment of wheat leaves with the *Bipolaris oryzae* (inducer organism) reduced spot blotch lesion numbers and sizes caused by *Bipolaris* on the same leaves (Sarhan et al. 1991). A preliminary study identified antifungal substances in apoplastic fluids of 'induced' plants that were able to prevent spore germination of *Bipolaris sorokiniana* in vitro (Chakraborty and Sinha 1984). Plant growth-promoting fungal isolates of Phoma spp. from Zoysia grass (Zoysia sp.) suppressed *Bipolaris sorokiniana* due to the competitive root colonization (Shivanna et al. 1996). Successful antagonists against seed-borne *Bipolaris* were Chaetomium sp., Idriella bolleyi, and *Gliocladium roseum* (Knudsen et al. 1995). Suppression of soil-borne fungi, including B. *sorokiniana*, has also been observed in the presence of isothiocyanates released into soil by Brassica species (Kirkegaard et al. 1996; Sarwar et al. 1998).

References

Acharya K, Dutta AK, Pradhan P (2011) "Bipolaris sorokiniana" (Sacc.) Shoem.: The most destructive wheat fungal pathogen in the warmer areas. 5:1064.

Adlakha KL, Wilcoxson RD, Ray-Chauduri SP (1984) Resistance of Wheat to Leaf Spot Caused by Bipolaris sorokiniana. Plant Dis 68:320–321.

Aggarwal R, Singh VB, Shukla R, Gurjar MS, Gupta S, Sharma TR (2010) URP-based DNA Fingerprinting of Bipolaris sorokiniana Isolates Causing Spot Blotch of Wheat. J Phytopathol 158:210–216. doi: 10.1111/j.1439-0434.2009.01603.x

Aggarwall R, Tewari AK, Srivastava KD, Singh D V (2004) Role of antibiosis in the biological control of spot blotch (Cochliobolus sativus) of wheat by Chaetomium globosum. Mycopathologia 157:369–377.

Alam KB, Shaheed MA, Ahmed AU, Malaker PK (1994) Bipolaris leaf blight (spot blotch) of wheat in Bangladesh. In: Saunders DA, Hettel GP (eds) Wheat in Heat-Stressed Environments Irrigated, Dry Areas and Rice-Wheat Farming Systems. CIMMYT, Mexico, D.E., pp 339–342

Alcorn JL (1988) The Taxonomy of "Helminthosporium" Species. Annu Rev Phytopathol 26:37–56. doi: 10.1146/annurev.py.26.090188.000345

Andres MF, Romero MD, Montes MJ, Delibes A (2001) Genetic relationships and isozyme variability in the Heterodera avenae complex determined by isoelectrofocusing. Plant Pathol 50:270–279.

Arun B, Joshi AK, Chand R, Singh BD (2003) Wheat somaclonal variants showing earliness, improved spot blotch resistance and higher yield. Euphytica 132:235–241. doi: 10.1023/A:1025097224408

Baichai W, Yugan L (1983) An investigation on the sources of primary inoculum of wheat CRR and a comparison of methods of counting condia in soil in Guangdong. J South China Agric Coll 4:88–96.

Bailey KL, Duczek LJ (1996) Managing cereal diseases under reduced tillage. Can J Plant Pathol 18:159–167.

Bakonyi J, Aponyi I, Fischl G (1998) Diseases caused by Bipolaris sorokiniana and Drechslera tritici repentis in Hungary. In: Duveiller E, Dubin HJ, Reeves J, McNab A (eds) Helminthosporium Blights of Wheat Spot Blotch and Tan Spot. Mexico, D.F., pp 80–85

Berestetskiy AO (2008) A review of fungal phytotoxins: from basic studies to practical use. Appl Biochem Microbiol 44:453–465. doi: 10.1134/S0003683808050013

Bhushan B, Singh K, Kaur S, Nanda GS (2002) Inheritance and allelic relationship of leaf blight resistance genes in three bread wheat varieties in the adult plant stage. J Genet Plant Breed 56:69–76.

Bisen BS, Channy B (1983) Some observations on the surface of wheat leaves during the early stages of infection by Helminthosporium sativum. J Indian Bot Soc 62:285–287.

Biswas SK, Srivastava KD, Aggarwal R, Prem D, Singh D V, Dureja P (2000) Antagonism of Chaetomium globosum to Drechslera sorokiniana, the spot blotch pathogen of wheat. Indian Phytopathol 53:436–440.

Bochow H, Kogel KH, Poehling H, von Alten H, Wyss U (2001) Mechanisms of interactions in the system of plant, parasite and beneficial organisms. J Plant Dis Prot 108:626–652.

Brandle JE, Namwilla JCP, Little R (1987) Effect of plant architecture on levels of Helminthosporium sativum infection on spring wheat grown in Zambia. Crop Prot 6:153–156.

Bunkina IA, Vasil'eva LN, Gornostai VI, Nazarova MM, Egorova LN, Zhukovskaya SA, Plyusnina RA (1978) Algae, fungi and mosses of the Soviet Far East (Vodorosli, griby i mkhi Dal'nego Vostoka). In: Azbukina ZM (ed) Vodorosli, griby i mkhi Dal'nego Vostoka. p 127.

Burdon JJ, Silk J (1997) Sources and patterns of diversity in plant pathogenic fungi. Phytopathology 87:664–669.

Chakraborty, S., & Newton, A. C. (2011). Climate change, plant diseases and food security: an overview. Plant Pathology, 60, 2–14. doi:10.1111/j.1365-3059.2010.02411.

Carlson H, Nilsson P, Jansson H-B, Odham G (1991) Characterization and determination of prehelminthosporol, a toxin from the plant pathogenic fungus Bipolaris sorokiniana, using liquid chromatography/mass spectrometry. J Microbiol Methods 13:259–269.

Cervone F, Castoria R, Leckie F, De Lorenzo G (1997) Perception of fungal elicitors and signal transduction. In: Aducci P (ed) Signal transduction in plants/: Molecular and Cell Biology Updates. Springer, pp 153–178.

Chakraborty D, Sinha AK (1984) Similarity between the chemically and biologically induced resistance in wheat seedlings to Drechslera sorokiniana. Zeitschrift für Pflanzenkrankheiten und Pflanzenschutz 91:59–64.

Chakraborty S, Perrott R, Ellis N, Thomas MR (1999) New aggressive Colletotrichum gloeosporioides strains on Stylosanthes scabra detected by virulence and DNA analysis. Plant Dis 83:333–340.

Chand R, Pandey SP, Singh H V, Kumar S, Joshi AK (2003) Variability and its probable cause in the natural populations of spot blotch pathogen Bipolaris sorokiniana of wheat (T. aestivum L.) in India. J Plant Dis Prot 110:27–35.

Chand R, Singh H V, Joshi AK, Duveiller E (2002) Physiological and Morphological Aspects of Bipolaris sorokiniana Conidia Surviving on Wheat Straw. Plant Pathol J 18:328–332.

Chavesa FC, Barbosa LCA, Demunera AJ, Silva AA (2006) New Helminthosporal Analogues with Plant-Growth Regulatory Properties Synthesized via Oxyallyl Cation. Zeitschrift für Naturforsch 61:1287–1294.

Chen X, Line RF, Leung H (1995) Relationship between virulence variation and DNA polymorphism in Puccinia striformis. Phytopathology 83:1489–1497. doi: 10.1094/Phyto-83-1489.

Chinkal P, Qiben C (1982) Studies on control of root rot caused by Bipolaris sorokiniana (Sacc. ex Sorokin) shoem. of spring wheat in North-Eastern China. Acta Phytophylacica Sin (Journal Plant Prot 4:1–10.

Chinn SHF, Ledingham RJ (1958) APPLICATION OF A NEW LABORATORY METHOD FOR THE DETERMINATION OF THE SURVIVAL OF HELMINTHOSPORIUM SATIVUM SPORES IN SOIL. Can J Bot 36:289–295.

Clark R V, Wallen VR (1969) Seed infection of barley by Cochliobolus sativus and its influence on yield. Can Plant Dis Surv 49:60–64.

Clay RP, Bergmann CW, Fuller MS (1997) Isolation and Characterization of an Endopolygalacturonase from Cochliobolus sativus and a Cytological Study of Fungal Penetration of Barley. Phytopathology 87:1148–59.

Coskun G, Zihnioglu F (2002) Effect of Some Biocides on Glutathione-s-transferase in Barley, Wheat, Lentil and Chickpea Plants. Turkish J Biol 26:89–94.

Couture L, Sutton JC (1978) Control of spot blotch in Barley by fungicides application timed according to weather factors. Phytoprotection 59:65–75.

Dastur JE (1942) Notes on some fungi isolated from "black point" affected kernels in the central provinces. Indian J Agric Sci 12731-742 12:731–742.

Dehne HW, Oerke EC (1985) Investigations on the occurrence of Cochliobolus sativus on barley and wheat. Vol. 2. Infection, colonization, and damage to stem and leaves. J Plant Dis Prot 92:606–617.

Dhaliwal HS, Singh H, Gill KS, Randhawa HS (1993) Evaluation and cataloguing of wheat germplasm for disease resistance and quality.

Dong N, Liu X, Lu Y, Du L, Xu H, Liu H, Xin Z, Zhang Z (2010) Overexpression ofTaPIEP1, a pathogen-induced ERF gene of wheat, confers host-enhanced resistance to fungal pathogen Bipolaris sorokiniana. Funct Integr Genomics 10:215–26. doi: 10.1007/s10142-009-0157-4.

Dubin HJ, Arun B, Begum SN, Bhatta M, Dhari R, Goel LB, Joshi AK, Khanna BM, Malaker PK, Pokhrel DR, Rahman MM, Saha NK, Shaheed MA, Sharma RC, Singh AK, Singh RM, Singh R V, Vergas M, Verma PC (1998) Result of South Asian regional Helminthosporium leaf blight and yield experiment 1993-1994. In: Duveiller E, Dubin HJ, Reeves J, McNab A (eds) Helminthosporium blights of wheat Spot blotch and tan spot. CIMMYT, Mexico D.F, pp 182–187.

Dubin HJ, Bimb HP (1994) Studies of soilborne diseases and foliar blights of wheat at the National Wheat Research Experiment Station, Bhairahawa, Nepal.

Dubin HJ, Rajaram S (1996) Breeding disease-resistant wheats for tropical highlands and lowlands. Annu Rev Phytopathol 34:503–526.

Dubin HJ, Van Ginkel M (1991) The status of wheat diseases and disease research in the warmer areas. In: Saunder DA, Hettel GP (eds) Wheat for Nontraditional Warm Areas in Heat Stressed Environments: Irrigated Dry Areas. UNDP/CIMMYT, Mexico, pp 125–145.

Duczek LJ (1981) Number and viability of conidia of Cochliobolus sativus in soil profiles in summer fallow fields in Saskatchewan. Can J Plant Pathol 3:12–14.

Duveiller E (2004) Controlling Foliar Blights of Wheat in the Rice-Wheat Systems of Asia. Plant Dis 88:552–556. doi: 10.1094/PDIS.2004.88.5.552

Duveiller E, Gilchrist LI (1994) Production constraints due to Bipolaris sorokiniana in wheat: Current situation and future prospects. In: Saunder DA, Hettel GP (eds) Wheat in Heat- Stressed Environments: Irrigated, Dry Areas and Rice-Wheat Farming Systems. CIMMYT, Mexico, D.F., pp 343–352.

Duveiller E, Kandel YR, Sharma RC, Shrestha SM (2005a) Epidemiology of foliar blights (spot blotch and tan spot) of wheat in the plains bordering the himalayas. Phytopathology 95:248–256.

Duveiller E, Sharma RC, Mercado D, Maraite H, Bhatta MR, Ortiz-Ferrara G, Sharma D (2005b) Controlling Foliar Blight of Wheat in South Asia/: A Holistic Approach. Turkish J Agric For 29:129–135.

Duveiller E, Singh RP, Nicol JM (2007) The challenges of maintaining wheat productivity: pests, diseases, and potential epidemics. Euphytica 157:417–430. doi: 10.1007/s10681-007-9380-z

Duveiller E, van Ginkel M, Dubin HJ (1998) Helminthosporium Diseases of Wheat/: Summary of Group Discussions and Recommendations. In: Duveiller E, Dubin HJ, Reeves J, McNab A (eds) Helminthosporium Blights of Wheat: Spot Blotch and Tan Spot. CIMMYT, Mexico D.F, p 376.

Duveiller EM, Sharma RC (2009) Genetic Improvement and Crop Management Strategies to Minimize Yield Losses in Warm Non-traditional Wheat Growing Areas due to Spot Blotch Pathogen Cochliobolus sativus. J Phytopathol 157:521–534. doi: 10.1111/j.1439-0434.2008.01534.x

Eyal Z, Blum A (1989) Canopy temperature as a correlative measure for assessing host response to Septoria tritici blotch of wheat. Plant Dis. 73:468–471.

Eyal Z, Scharen AL, Prescott JM, van Ginkel M (1987) The Septoria Diseases of Wheat: Concepts & Methods of Disease Management. CIMMYT, Mexico D.F.

Fetch TG, Steffenson BJ (1994) Identification of Cochliobolus sativus isolates expressing differential virulence on two-row barley genotypes from North Dakota. Can J Plant Pathol 16:202–206. doi: 10.1080/07060669409500754

Garrett, K. A., Dendy, S. P., Frank, E. E., Rouse, M. N., & Travers, S. E. (2006). Climate change effects on plant disease: genomes to ecosystems. Annual Review of Phytopathology, 44, 489–509. doi:10.1146/annurev.phyto.44.070505.143420.

Ghini, R., Bettiol, W., & Hamada, E. (2011a). Diseases in tropical and plantation crops as affected by climate changes: current knowledge and perspectives. Plant Pathology, 60, 122–132. doi:10.1111/j.1365-059.2010.02403.x.

Gayad SK (1961) Production of symptoms of barley leaf spot disease by cultural filtrates of Helminthosporium sativum. Nature 191:725–726.

Guzmán P, Gepts P, Temple S, Mkandawire ABC, Gilbertson RL (1999) Detection and Differentiation of Phaeoisariopsis griseola Isolates with the Polymerase Chain Reaction and Group-Specific Primers. Plant Dis 83:37–42. doi: 10.1094/PDIS.1999.83.1.37

Hait GN, Sinha AK (1986) Protection of wheat seedlings from Helminthosporium infection by seed treatment with chemicals. J Phytopathol 115:97–107.

Hawksworth DL, Kirk PM, Sutton BC, Pegler DN (1995) Ainsworth and Bisby's Dictionary of the Fungi (8th Ed.). CAB Int Wallingford, United Kingdom pp. 616.

Hetzler J, Eyal Z, Mehta YR (1991) Interaction between spot blotch (Cochliobolus sativus) and wheat cultivars. In: Saunders D (ed) Wheat for non-traditional warmer areas. CIMMYT, Mexico, D.F., pp 141–164.

Hobbs PR, Giri GS (1997) Reduced and zero tillage options for establishment of wheat after rice in South Asia. In: Braun HJ (ed) Wheat: prospects for global improvement: proceeding of the 5th International Wheat Conference, 10-14 June, 1996, Ankara, Turkey. Kluwer Academic Publishers, Dordrecht, Netherlands, pp 455–465.

Ibeagha AE, Hückelhoven R, Schäfer P, Singh DP, Kogel K-H (2005) Model Wheat Genotypes as Tools to Uncover Effective Defense Mechanisms Against the Hemibiotrophic Fungus Bipolaris sorokiniana. Phytopathology 95:528–532. doi: 10.1094/PHYTO-95-0528.

Janni M, Sella L, Favaron F, Blechl AE, De Lorenzo G, D'Ovidio R (2008) The expression of a bean PGIP in transgenic wheat confers increased resistance to the fungal pathogen Bipolaris sorokiniana. Mol Plant Microbe Interact 21:171–7. doi: 10.1094/MPMI-21-2-0171.

Jones DA, Jones JDG (1997) The role of leucine-rich repeat proteins in plant defences. Adv Bot Res 24:89–167.

Joshi AK, Chand R (2002) Variation and inheritance of leaf angle, and its association with spot blotch (Bipolaris sorokiniana) severity in wheat (Triticum aestivum). Euphytica 124:283–29.

Joshi AK, Chand R, Arun B (2002) Relationship of plant height and days to maturity with resistance to spot blotch in wheat. Euphytica 123:221–228.

Joshi AK, Chand R, Kumar S, Singh RP (2004a) Leaf Tip Necrosis: A Phenotypic Marker Associated with Resistance to Spot Blotch Disease in Wheat. Crop Sci 44:792–796.

Joshi AK, Kumar S, Chand R, Ortiz-Ferrara G (2004b) Inheritance of resistance to spot blotch caused by Bipolaris sorokiniana in spring wheat. Plant Breed 123:213–219. doi: 10.1111/j.1439-0523.2004.00954.x

Joshi AK, Kumari M, andC. M. Reddy VPS, Kumar S, Rane J, Chand R (2007a) Stay green trait: variation, inheritance and its association with spot blotch resistance in spring wheat (Triticum aestivum L.). Euphytica 153:59–71.

Joshi AK, Mishra B, Chatrath R, Ortiz Ferrara G, Singh RP (2007b) Wheat improvement in India: present status, emerging challenges and future prospects. Euphytica 157:431–446. doi: 10.1007/s10681-007-9385-7.

Joshi AK, Ortiz-Ferrara G, Crossa J, Singh G, Alvarado G, Bhatta MR, Duveiller E, Sharma RC, Pandit DB, Siddique AB, Das SY, Sharma RN, Chand R

(2007c) Associations of Environments in South Asia Based on Spot Blotch Disease of Wheat Caused by. Crop Sci 47:1071. doi: 10.2135/cropsci 2006.07.0477.

Joshi AK, Ortiz-Ferrara G, Crossa J, Singh G, Sharma RC, Chand R, Parsad R (2007d) Combining superior agronomic performance and terminal heat tolerance with resistance to spot blotch (Bipolaris sorokiniana) of wheat in the warm humid Gangetic Plains of South Asia. F Crop Res 103:53–61.

Joshi LM (1986) Monitoring of wheat disease during 1984-85. Indian J Mycol Plant Pathol 15:135–139.

Juroszek, P., & von Tiedemann, A. (2011). Potential strategies and future requirements for plant disease management under a changing climate. Plant Pathology, 60, 100–112. doi:10.1111/j.1365-3059.2010.02410.

Kemp G, Bergmann CW, Clay R, Van der Westhuizen AJ, Pretorius ZA (2003) Isolation of a polygalacturonase-inhibiting protein (PGIP) from wheat. Mol Plant Microbe Interact 16:955–61. doi: 10.1094/MPMI.2003.16.11.955.

Khan H, Chowdhury S (2011) Identification of resistance source in wheat germplasm against spot blotch disease caused by Bipolaris sorokiniana. Arch Phytopathol Plant Prot 44:840–844. doi: 10.1080/0323540 0903345273.

Khasanov BA (1988) Yellow leaf spot of cereals caused by Pyrenophora-tritici-repentis (Died.) Drechs. Mikol i Fitopatol 22:78–83.

Kirkegaard JA, Wong PTW, Desmarchelier JM (1996) In vitro suppression of fungal root pathogens of cereals by Brassica tissues. Plant Pathol 45:593–603.

Kline DM, Nelson RR (1963) Pathogenicity of isolates of Cochliobolus sativus from cultivated and wild gramineous hosts from the Western Hemisphere to species of the Gramineae. Plant Dis Report 47:890–894.

Knudsen IMB, Hockenhull J, Jensen DF (1995) Biocontrol of seedling disease of barley and wheat caused by Fusarium culmorum and Bipolaris sorokiniana: Effects of selected fungal antagonists on growth and yield components. Plant Pathol 44:467–477.

Kohli MM, Mann CE, Rajaram S (1991) Global status and recent progress in breeding wheat for the warmer areas. In: Saunders DA (ed) Wheat for Non-traditional Warm Areas. CIMMYT, Mexico, D.F., pp 96–112.

Kumar D, Chand R, Prasad LC, Joshi AK (2007a) A new technique for monoconidial culture of the most aggressive isolate in a given population of Bipolaris sorokiniana, cause of foliar spot blotch in wheat and barley. WORLD J Microbiol Biotechnol 23:1647–1651.

Kumar J, Schäfer P, Hückelhoven R, Langen G, Baltruschat H, Stein E, Nagarajan S, Kogel K-H (2002) Bipolaris sorokiniana, a cereal pathogen of global concern: cytological and molecular approaches towards better controldouble dagger. Mol Plant Pathol 3:185–195.

Kumar S, Prasad LC, Kumar U, Tyagi K, Arun B, Joshi AK (2007b) Inheritance and allelic relationship of resistance genes to spot blotch of wheat caused by Bipolaris sorokiniana. In: Buck HT et al (ed) Wheat Production in Stressed Environments. Springer., pp 113–118.

Kumar S, Roeder MS, Tripathi SB, Kumar S, Chand R, Joshi AK, Kumar U, 2015. Mendelization and fine mapping of a bread wheat spot blotch disease resistance QTL. Molecular Breeding 35, 218.

Kumar S, Roeder MS, Singh RP, Chand R, Joshi AK, Kumar U (2016). Mapping of spot blotch disease resistance using NDVI as a substitute to visual observation in wheat (Triticum aestivum L.). Molecular Breeding 36, 95.

Kumar U, Joshi AK, Kumar S, Chand R, Röder MS (2010) Quantitative trait loci for resistance to spot blotch caused by Bipolaris sorokiniana in wheat (T. aestivum L.) lines "Ning 8201" and "Chirya 3." Mol Breed 26:477–491. doi: 10.1007/s11032-009-9388-2.

Kumar U, Joshi AK, Kumar S, Chand R, Röder MS (2009) Mapping of resistance to spot blotch disease caused by Bipolaris sorokiniana in spring wheat. Theor Appl Genet 118:783–792.

Kumar U, Kumar S, Tyagi K, Chand R, Joshi AK (2005) Microsatellite markers for resistance to spot blotch in spring wheat. Commun Agric Appl Biol Sci 70:59–60.

Kumar U, Tyagi K, Chand R, Joshi AK, Kumar S (2004) Identification of RAPD marker linked to spot blotch resistance gene using bulked Segregant Analysis in wheat (T. aestivum). 74th annual session of National Academy of Science India. NASI Allahabad India, pp 9–10.

Lagudah ES, Krattinger SG, Herrera-Foessel SA, Singh RP, Huerta-Espino J, Spielmeyer W, Brown-Guedira G, Selter LL, Keller B (2009) Gene-specific markers for the wheat gene Lr34/Yr18/Pm38 which confers resistance to multiple fungal pathogens. Theor Appl Genet 119:889–898.

Ledingham RJ, Sallans BJ, Wenhardt A (1960) INFLUENCE OF CULTURAL PRACTICES ON INCIDENCE OF COMMON ROOTROT IN WHEAT IN SASKATCHEWAN. Can J Plant Sci 40:310–316. doi: 10.4141/cjps60-042.

Lespinay DA (2004) Selection for stable resistance to Helminthosporium leaf blights in non-traditional warm wheat areas. Université Catholique de Louvain; Louvain-La-Neuve; Belgium.

Liljeroth E, Franzon-Almgren I, Gustafsson M (1994) Effect of prehelminthosporol, a phytotoxin produced by Bipolaris sorokiniana, on barley roots. Can J Bot 72:558–563.

Lillemo M, Joshi AK, Prasad R, Chand R, Singh RP (2013) QTL for spot blotch resistance in bread wheat line Saar co-locate to the biotrophic disease resistance loci Lr34 and Lr46. Theor Appl Genet 126:711–719. doi: 10.1007/s00122-012-2012-6.

Lu P, Liang Y, Li D et al., 2016. Fine genetic mapping of spot blotch resistance gene Sb3 in wheat (Triticum aestivum). Theoretical and Applied Genetics 129, 577–89.

Ludwig RA (1957) Toxin production by Helminthosporium sativum P.K. & B. and its significance in disease development. Can J Bot 35:291–303.

Luz WC Da, Bergstrom GC (1986) Temperature-sensitive development of spot blotch in spring wheat cultivars differing in resistance. Fitopatol Bras 11:197–204.

Mahto BN, Gurung S, Adhikari TB (2011) Assessing genetic resistance to spot blotch, Stagonospora nodorum blotch and tan spot in wheat from Nepal. Eur J Plant Pathol 131:249–260. doi: 10.1007/s10658-011-9803-5.

Malaker PK, Mian LH, Khandaker MDM, Reza MMA (2008) Survival of Bipolaris sorokiniana (Sacc.) Shoemaker in soil and residue of wheat. Bangladesh J Bot 36:133–137. doi: 10.3329/bjb.v36i2.1501.

Malvick DK, Grau CR (2001) Characteristics and Frequency of Aphanomyces euteiches Races 1 and 2 Associated with Alfalfa in the Midwestern United States. Plant Dis 85:740–744. doi: 10.1094/PDIS.2001.85.7.740.

Manamgoda SD, Cai L, Bahkali AH, Chukeatirote E, Hyde DK (2011) Cochliobolus: an overview and current status of species. Fungal Divers 51:3–42. doi: 10.1007/s13225-011-0139-4.

Mandal S, Srivastava K, Aggarwal R, Singh D (1999) Mycoparasitic action of some fungi on spot blotch pathogen (Drechslera sorokiniana) of wheat. Indian Phytopathol. 52:39–43.

Maraite H, Di Zinno T, Longree H, Daumerie V, Duveiller E (1998) Fungi association with foliar blight of wheat in warm areas. In: Duvellier E, Dubin HJ, Reeves J, McNab A (eds) Helminthosporium blights of wheat spot blotch and tan spot. CIMMYT, Mexico, D.F., pp 293–300.

Mayo P De, Williams RE, Spencer EY (1965) TERPENOIDS: VIII. THE IMMEDIATE PRECURSORS OF HELMINTHOSPORAL AND HELMINTHOSPOROL. Can J Chem 43:1357–1365. doi: 10.1139/v65-184.

McDermott JM, McDonald BA (1993) Gene flow in plant pathosystems. Annu Rev Phytopathol 31:353–373.

McIntosh RA (2007) Close genetic linkage of genes conferring adult-plant resistance to leaf rust and stripe rust in wheat. Plant Pathol 41:523–527. doi: 10.1111/j.1365-3059.1992.tb02450.x

Mehta YR (1993) Spot blotch (Bipolaris sorokiniana). In: Mathur SB, Cunfer BM (eds) Seed borne disease and seed health testing of wheat. Institute of seed pathology for developing countries, Copenhagen, Denmark, pp 105–112.

Mehta YR (1994) Manejo Integrado de Enfermedadas de Trigo-santa Cruz. Bolivia. CIAT/IAPAR pp. 314.

Mehta YR (1985) Breeding wheat for resistence to Helminthosprium spot blotch. In: Wheat for the tropical Environment. A proceeding of The International symposium. CIMMYT, Mexico, D.F., pp 135–144.

Mercado D, Renard ME, Maraite H, Duveiller E (2003) Chlorophyll content and chlorophyll fluorescence as indicators of resilience to temperature stress in wheat and its relationship with resistance to Bipolaris sorokiniana. Proc. Fourth Int. Wheat Tan Spot Spot Blotch Work. Bemidji Minnesota USA 2124 July 2002.

Melloy P, Hollaway G, Luck J, Norton R, Aitken E And Chakraborty S, (2010) Production and fitness of Fusarium pseudograminearum inoculum at elevated carbon dioxide in FACE 2010 DOI: 0.1111/j.1365-2486. 2010.02178.

Mikhailova LA, Lianfa S, Gogoleva SG, Gultyaeva EI (2004) Sources and donors of resistance to wheat spot blotch caused by Bipolaris sorokiniana Shoem. (Cochliobolus sativus Drechs. ex Dastur). Arch Phytopathol Plant Prot 37:161–167. doi: 10.1080/03235400310001631909.

Millano D, Zadoks JC (1985) The effect of early foliar infection by Helminthosporium sativum on some yield components of two African wheats. A Proceedings of the International Symposium: Wheats for More Tropical Environments. CIMMYT, Mexico, DF. pp 154–157.

Mishra AB, Chourasia RK (1976) Pathological and epidemiological studies on leaf blight of wheat (Helminthosporium sativum) Pamm. & King. Mysore J Agric Sci 10:279–285.

Mishra AP (1981) Variability; physiologic specialization and genetics of pathogenicity in graminicolus Helminthosporium affecting cereal crops. Indian Phytopathol 34:1–22.

Misra AP (1973) Helminthosporium species occurring on cereals and other Gramineae. U.S.P.L. 480 Project No. A7-CR 133, Grant No. FG-IN-223. Tirhut College of Agriculture, Dholi, Muzaffarpur, Bihar, India. Catholic Press, Ranchi, Bihar. pp. 289.

Mokabli A, Valette S, et Rivoal R (2001) Differentiation of some species of cereal and graminaceous cyst nematodes by cellulose acetate electrophoresis. Nematol Mediterr 29:103–108.

Mujeeb-Kazi A (2003) Wheat Improvement Facilitated by Novel Genetic Diversity and In vitro Technology. Plant Tissue Cultture 13:179–210.

Mujeeb-Kazi A, Gul A, Farooq M, Rizwan S, Mirza JI (2007) Genetic diversity of Aegilops variabilis (2n=4x=28; UUSS) for wheat improvement: Morpho-cytogenetic characterization of some derived amphiploids and their practical significance. Pakistan J Bot 39:57–66.

Mujeeb-Kazi A, Rosas V, Roldan S (1996a) Conservation of the genetic variation of Triticum tauschii (Coss.) Schmalh. (Aegilops squarrosa auct. non L.) in synthetic hexaploid wheats (T. turgidum L. s.lat. x T. tauschii; 2n=6x=42,

AABBDD) and its potential utilization for wheat improvement. Genet Resour Crop Evol 43:129–134.

Mujeeb-Kazi A, Villareal RL, Gilchrist LA, Rajaram S (1996b) Registration of five wheat germplasm lines resistant to Helminthosporium leaf blight. Crop Sci 36:216–217.

Nagarajan S, Kumar J (1998) Foliar blights of wheat in Indian germplasm: Improvement and future challenges for sustainable high yielding wheat production. In: Duveiller E, Dubin HJ, Reeves J, Mcnab A (eds) Proceedings on International Workshop on Helminthosporium Diseases of Wheat Spot Blotch and Tan Spot. CIMMYT, Mexico, DF . Feb 9-14. pp 52–58.

Neergaard P (1997) Seed Pathology, 1st edn. Palgrave Macmillan.

Nema KG, Joshi LH (1973) Spot blotch disease of wheat in relation to host age, temperature and moisture. Indian Phytopathol 26:41–48.

Neupane AC, Sharma RC, Duveiller E, Shrestha SM (2010) Sources of Cochliobolus sativus Inoculum Causing Spot Blotch under Warm Wheat Growing Conditions in South Asia. Cereal Res Commun 38:541–549. doi: 10.1556/CRC.38.2010.4.11.

Neupane RB, Sharma RC, Duveiller E, Ortiz-Ferrara G, Ojha BR, Rosyara UR, Bhandari D, Bhatta MR (2007) Major Gene Controls of Field Resistance to Spot Blotch in Wheat Genotypes “Milan/Shanghai #7” and “Chirya.3.” Plant Dis 91:692–697.

Oliveira AMR, Matsumura ATS, Prestes AM, Matos GS, Van Der Sand ST (1998) Variabilidade patogênica e morfológica em isolados de Bipolaris sorokiniana. Fitopatol Bras (Tropical Plant Pathol 23:349–353.

Oliveira AMR de, Matsumura ATS, Prestes AM, Van Der Sand ST (2002) Intraspecific variability of Bipolaris sorokiniana isolates determined by random-amplified polymorphic DNA (RAPD). Genet Mol Res 1:350–358.

Ortiz R, Braun H-J, Crossa J, Crouch JH, Davenport G, Dixon J, Dreisigacker S, Duveiller E, He Z, Huerta J, Joshi AK, Kishii M, Kosina P, Manes Y, Mezzalama M, Morgounov A, Murakami J, Nicol J, Ferrara GO, Ortiz-Monasterio JI, Payne TS, Peña RJ, Reynolds MP, Sayre KD, Sharma RC, Singh RP, Wang J, Warburton M, Wu H, Iwanaga M (2008) Wheat genetic resources enhancement by the International Maize and Wheat Improvement Center (CIMMYT). Genet Resour Crop Evol 55:1095–1140. doi: 10.1007/s10722-008-9372-4.

P Sharma EDRCS, Sharma P, Duveiller E, Sharma RC (2006) Effect of mineral nutrients on spot blotch severity in wheat, and associated increases in grain yield. F Crop Res 95:426–430. doi: 10.1016/j.fcr.2005.04.015.

Pammel LH, King CM, Bakke AL (1910) Two barley blights, with comparison of species of Helminthosporium upon cereals. Bull Iowa State Coll 116:178–190.

Pandey SP, Kumar S, Kumar U, Chand R, Joshi AK (2005) Sources of inoculum and reappearance of spot blotch of wheat in rice–wheat cropping systems in eastern India. Eur J Plant Pathol 111:47–55.

Pandey SP, Sharma S, Chand R, Shahi P, Joshi AK (2008) Clonal variability and its relevance in generation of new pathotypes in the spot blotch pathogen, Bipolaris sorokiniana. Curr Microbiol 56:33–41.

Patil L C, Hanchinal RR (2011) Identification of RAPD markers for spot blotch resistance in wheat. Karnataka J. Agric. Sci. 24:

Pederson VD, Nutter FWJ (1984) Estimation of Barley Yield Loss from Spot Blotch using Remote Sensing Techniques. Barley Newsl 27:49.

Pena-Rodriguez LM, Chilton WS (1989) Victoxinine and Prehelminthosporolactone, Two Minor Phytotoxic Metabolites Produced by Bipolaris sp., a Pathogen of Johnson Grass. J Nat Prod 52:899–901. doi: 10.1021/np50064a046.

Pierto DA (1990) Studies on the biology of Chaetomium globosum Kunze and its mode of action as an antagonist of Pythium ultimum Trow. University of Basel, Switzerland.

Poloni A, Muller MV, Pessi I, Sand ST Van Der (2008) Analysis of morphological and growth rate variability of polyconidial and monoconidial cultures of Bipolaris sorokiniana. Biociências(On-line) 16:

Poloni A, Pessi IS, Frazzon APG, Van der Sand ST (2009) Morphology, physiology, and virulence of Bipolaris sorokiniana isolates. Curr Microbiol 59:267–73. doi: 10.1007/s00284-009-9429-4

Prescott JM, Burnett PA, Saari EE, Ranson J, Bowman J, Milliano W de, Singh RP, Bekele G (1986) Wheat Diseases and Pests: a guide for field identification: CIMMYT, Mexico, D.F.

Qixiang M, Jiabi H (1987) On biological study of a wheat leaf spot pathogen Bipolaris sorokiniana. Acta Phytopathol Sin 17:59–64.

Raemaekers R. (1991) Contribution to the epidemiology of Bipolaris sorokiniana diseases and the development of rainfed wheat, a new crop in Zambia. Katholieke Universiteit te Leuven, Belgium.

Raemaekers R. (1988) Helminthosporium sativum: Disease complex on wheat and sources of resistance in Zambia. In: Klatt AR (ed) In: Wheat Production Constraints in Tropical Environments. CIMMYT, Mexico, D.F., pp. 175–185.

Ragiba M, Prabhu K V (2009) Inheritance of resistance to spot blotch disease in wheat. Indian J Genet Plant Breed 69: 178–182.

Ragiba M, Prabhu K V, Singh RB (2004) Recessive genes controlling resistance to Helminthosporium leaf blight in synthetic hexaploid wheat. Plant Breed 123:389–391.

Reis EM (1991) Integrated disease management- The changing concept of controlling head blight and spot blotch. In: Saunders DA, Hettel GP (eds.)

Wheat in Heat-Stressed Environments Irrigated, Dry Areas and Rice-Wheat Farming Systems. CIMMYT, Mexico, D.F., pp 165–177.

Reis EM, Forcelini CA (1994) Transmissao de Bipolaris sorokiniana de sementes para orgaos radiculares e aereos do trigo. Fitopatol Bras (Tropical Plant Pathol 18:76–81.

Reis EM, Medeiros CA, Casa TA (1998) Control of leaf blight of wheat by elimination of inoculum source. In: Duveiller E, Dubin HJ, Reeves J (eds) Helminthosporium Blights of Wheat: Spot Blotch and Tan Spot. CIMMYT, Mexico, D.F., pp 327–332.

Reis EM, Santos HP (1987) The increased sporulation of Cochliobolus sativus on above ground tissues of small grains and its relationship to the origin of inoculum in the soil. Fitopatol Bras (Tropical Plant Pathol 12:206–208.

Reis EM, Wünsche WA (1984) Sporulation of Cochliobolus sativus on Residues of Winter Crops and Its Relationship to the Increase of Inoculum Density in Soil. Plant Dis 68:411–412.

Ridley BL, O'Neill MA, Mohnen D (2001) Pectins: structure, biosynthesis, and oligogalacturonide-related signaling. Phytochemistry 57:929–67.

Rosenow DT (1984) Sorghum root and stalk rots, a critical review. In: Mughogho LK, Rosenberg G (eds) Breeding for resistance to root and stalk rots in Texas. ICRISAT, Hyderabad, India, pp 209–217.

Rosewarne GM, Singh RP, Huerta-Espino J, William HM, Bouchet S, Cloutier S, McFadden H, Lagudah ES (2006) Leaf tip necrosis, molecular markers and beta1-proteasome subunits associated with the slow rusting resistance genes Lr46/Yr29. Theor Appl Genet 112:500–8. doi: 10.1007/s00122-005-0153-6.

Rosyara UR, Khadka K, Subedi S, Sharma RC, Duveiller E, Rosyara U R Khadka K Subedi S Sharma R C Duveiller E (2009) Field resistance to spot blotch is not associated with undesirable physio-morphological traits in three spring wheat populations. J Plant Pathol 91:113–122. doi: 10.4454/jpp.v91i1.631.

Rosyara UR, Subedi S, Duveiller E, Sharma RC (2010) The effect of spot blotch and heat stress on variation of canopy temperature depression, chlorophyll fluorescence and chlorophyll content of hexaploid wheat genotypes. Euphytica 174:377–390. doi: 10.1007/s10681-010-0136-9.

Saari EE (1998) Leaf blight disease and associated soil borne fungal pathogens of wheat in South and South East Asia. In: Duveiller E, Dubin HJ, Reeves J, McNab A (eds) Helminthosporium Blights of Wheat: Spot Blotch and Tan Spot. CIMMYT, Mexico, D.F., pp 37–51.

Saari EE, Prescott JM (1975) A scale for appraising the foliar intensity of wheat diseases. Plant Dis Reporter1 59:377–380.

Sarhan ART, Király Z, Sziráki I, Smedegaard-Petersen V (1991) Increased levels of cytokinins in barley leaves having the systemic acquired resistance to Bipolaris Sorokiniana (Sacc.) Shoemaker. J Phytopathol 131:101–108.

Sarwar M, Kirkegaard JA, Ong PTW, Desmarchelier J. (1998) Biofumigation potential of brassica-III. In vitro toxicity of isothiocyanates to soil-borne fungal pathogens. Plant Soil 201:103–112.

Saunders DA (1988) Characterization of tropical wheat environments: Identification of production constraints and progress achieved in South and South East Asia. In: Klatt AR (ed) Wheat Production Constraints in Tropical Environments. CIMMYT, Mexico, D.F., pp 12–26.

Sentelhas PC, Pedro Júnior MJ, Felício JC (1993) Effects of different conditions of irrigation and crop density in microclimate and spot blotch and powdery mildew occurrence. Bragantia 52:45–52. doi: 10.1590/S0006-87051993000100006.

Shaner G (1981) Effect of Environment on Fungal Leaf Blights of Small Grains. Annu Rev Phytopathol 19:273–296. doi: 10.1146/annurev.py.19.090181.001421.

Sharma HC, Crouch JH, Sharma KK, Seetharama N, Hash CT (2002) Applications of biotechnology for crop improvement: prospects and constraints. Plant Sci 163:381–395.

Sharma RC, Bhatta MR (1999) Independent inheritance of maturity and spot blotch resistance in wheat. J Inst Agric Anim Sci 19-20:111–118.

Sharma RC, Dubin HJ (1996) Effect of wheat cultivar mixtures on spot blotch (Bipolaris sorokiniana) and grain yield. F Crop Res 48:95–101. doi: 10.1016/S0378-4290(96)01031-3.

Sharma RC, Dubin HJ, Bhatta MR, Devkota RN (1997a) Selection for spot blotch resistance in four spring wheat populations. Crop Sci 37:432–435.

Sharma RC, Dubin JH, Devkota RN, Bhatta M. (1997b) Heritability estimates of field resistance to spot blotch in four spring wheat crosses. Plant Breed 116:64–68.

Sharma RC, Duveiller E (2007) Advancement toward New Spot Blotch Resistant Wheats in South Asia. Crop Sci 47:961–968. doi: 10.2135/cropsci 2006.03.0201.

Sharma RC, Duveiller E, Ahmed F, Arun B, Bhandari D, Bhatta MR, Chand R, Chaurasiya PCP, Gharti B, Hossain MH, Joshi AK, Mahto BN, Malaker PK, Reza MA, Rahman M, Samad MA, Shaheed MA, Siddique AB, Singh AK, Singh KP, Singh RN, Singh S. (2004) Helminthosporium leaf blight resistance and agronomic performance of wheat genotypes across warm regions of South Asia. Plant Breed 123:520–524.

Sharma RC, Duveiller E, Jacquemin JM (2007a) Microsatellite Markers Associated with Spot Blotch Resistance in Spring Wheat. J Phytopathol 155:316–319. doi: 10.1111/j.1439-0434.2007.01238.x

Sharma RC, Duveiller E, Ortiz-Ferrara G (2007b) Progress and challenge towards reducing wheat spot blotch threat in the Eastern Gangetic Plains of South Asia: Is climate change already taking its toll? F Crop Res 103:109–118. doi: 10.1016/j.fcr.2007.05.004

Sharma RC, Ortiz-Ferrara G, Crossa J, Bhatta MR, Sufian MA, Shoran J, Joshi AK, Chand R, Singh G, Ortiz R (2007c) Wheat grain yield and stability assessed through regional trials in the eastern Gangetic plains of south Asia. Euphytica 157:457–464.

Sharma RC, Pandey-Chhetri B, Duveiller E (2006) Heritability estimates of spot blotch resistance and its association with other traits in spring wheat crosses. Euphytica 147:317–327. doi: 10.1007/s10681-005-9018-y.

Sharma RC, Tiwary AK, Ortiz-Ferrara G (2008) Reduction in kernel weight as a potential indirect selection criterion for wheat grain yield under terminal heat stress. Plant Breed 127:241–248. doi: 10.1111/j.1439-0523.2007.01460.x

Sharma-Poudyal D, Duveiller E, Sharma RC (2005) Effects of Seed Treatment and Foliar Fungicides on Helminthosporium Leaf Blight and on Performance of Wheat in Warmer Growing Conditions. J Phytopathol 153:401–408. doi: 10.1111/j.1439-0434.2005.00992.x

Shivanna MB, Meera MS, Kageyama K, Hyakumachi M (1996) Growth promotion ability of zoysiagrass rhizosphere fungi in consecutive plantings of wheat and soybean. Mycoscience 37:163–168.

Shoemaker RA (1959) Nomenclature of Drechslera and Bipolaris, grass parasites segregated from Helminthosporium. Can J Bot 37:879–887.

Siddique AB, Hossain MH, Duveiller E, Sharma RC (2006) Progress in Wheat Resistance to Spot Blotch in Bangladesh. J Phytopathol 154:16–22. doi: 10.1111/j.1439-0434.2005.01049.x

Singh BN, Singh RN, Singh AK, Singh SP (2000) Inheritance of resistance in wheat to Cochliobolus sativus causing spot blotch. Indian Phytopathol 53:486–487.

Singh D, Singh RV, Singh AK, Singh BN (1998a) Identification and inheritance of resistance to the foliar blight of wheat. In: Duveiller E, Dubin HJ, Reeves J, McNab A (eds) Helminthosporium Blights of Wheat: Spot Blotch and Tan Spot/: Proceedings of an International Workshop; CIMMYT, Mexico, 9-14 February. CIMMYT, Mexico, D.F., p 2.

Singh PJ, Dhaliwal HS (1993) Resistence to foliar blight of wheat in Agilops and wild Triticum spesies. Indian Phytopathol 46:246–248.

Singh RP (1992a) Genetic Association of Leaf Rust Resistance Gene Lr34 with Adult Plant Resistance to Stripe Rust in Bread Wheat. Phytopathology 82:835–838. doi: 10.1094/Phyto-82-835.

Singh RP (1992b) Association between Gene Lr34 for Leaf Rust Resistance and Leaf Tip Necrosis in Wheat. Crop Sci 32:874. doi: 10.2135/cropsci1992.0011183X003200040008x.

Singh RP, Ma H, Rajaram S (1995) Genetic Analysis of Resistance to Scab in Spring Wheat Cultivar Frontana. Plant Dis 79:238–240.

Singh R V, Singh AK, Ahmed R, Singh SP (1998b) Influence of agronomic practices on foliar blight, and identification of alternate hosts in rice-wheat cropping system. In: Duveiller E, Dubin HJ, Reeves J, McNab A (eds) Helminthosporium Blights of Wheat: Spot Blotch and Tan Spot/: Proceedings of an International Workshop; CIMMYT, Mexico, 9-14 February. pp 333–338.

Sivanesan A (1987) Mycological Papers Vol. 158: Graminicolous Species of Bipolaris, Curvularia, Drechslera, Exserohilum & Their Teleomorphs. C.A.B. International Mycological Institute.

Smurova SG, Mikhailova LA (2007) Sources of resistance to wheat spot blotch. Russ Agric Sci 33:378–380. doi: 10.3103/S1068367407060092.

Srivastava OP, Luthra JK, Narula PN (1971) Inheritance of seedling resistance to leafblight of wheat. Indian J Genet Plant Breed 31:209–211.

Steffenson BJ, Hayes PM, Kleinhofs A (,1996) Genetics of seedling and adult plant resistance to net blotch (Pyrenophora teres f. teres) and spot blotch (Cochliobolus sativus) in barley. Theor Appl Genet 92:552–558. doi: 10.1007/BF00224557.

Sticher L, Mauch-Mani B, Métraux JP (1997) Systemic acquired resistance.

Subramanian C V., Jain BL (1966) A Revision of Some Graminicolous Helminthosporia. Curr Sci 35:352–355.

Tamura S, Sakurai A, Kainuma K, Takai M (1963) Isolation of helminthosporol as a natural plant growth regulator and its chemical structure. Agric Biol Chem 27:738–739.

Tamura S, Sakurai A, Kainuma K, Takai M (1965) Isolation of helminthosporol as a natural plant growth regulator and its chemical structure. Agric Biol Chem 29:216–221.

The TT, Gupta RB, Dyck PL, Appels R, Hohmann U, McIntosh RA (1991) Characterization of stem rust resistant derivatives of wheat cultivar Amigo.

Tinline RD (1962) Cochliobolus sativus. V. Heterocariosis and parasexuality. Can J Bot 40:425–437.

Tinline RD (1951) Studies on the perfect stage of Helminthosporium sativum. Can J Bot 29:467–478. doi: 10.1139/b51-043.

Tinline RD, Dickson JG (1958) Cochliobolus sativus. I. Perithecial Development and the Inheritance of Spore Color and Mating Type. Mycologia 50:697–706.

Valjavec-Gratian M, Steffenson BJ (1997) Genetics of Virulence in Cochliobolus sativus and Resistance in Barley. Phytopathology 87:1140–1143. doi: 10.1094/PHYTO.1997.87.11.1140.

Van der Plank JE (1963) Plant diseases: Epidemics and control. Academic Press, University of Minnesota.

Van Ginkel M, Rajaram S (1998) Breeding for resistance to spot blotch in wheat Global perspective. In: Duveiller E, Dubin HJ, Reeves J, McNab A (eds)

Helminthosporium Blights of Wheat Spot Blotch and Tan Spot. CIMMYT, Mexico, D.F., pp 162–169.

Van Ginkel M, Rajaram S (1993) Breeding for durable resistance to diseases in wheat an additional perspective. In: Jacobes T, Parlevliet JE (eds) Durability of disease resistance. Kluwer Academic Publishers; Dordrecht; Netherlands, pp 259–272.

Vannacci G, Harman GE (1987) Biocontrol of seed-borne Alternaria raphani and A . brassicicola. Can J Microbiol 33:850–856. doi: 10.1139/m87-149.

Velazquez-Cruz C (1994) Genetica de la resistencia a Bipolaris sorokiniana en trigos harineros. Montecillos, Mexico.

Villareal LR, Mujeeb-Kazi A, Gilchrist IL, Toro E Del (1995) Yield loss to spot blotch in spring bread wheat in warm non-traditional wheat production areas. Plant Dis 79:893–897.

Villareal RL, Mujeeb-Kazi A, Fuentes Davila G, Rajaram S (1996) Registration of Four Synthetic Hexaploid Wheat Germplasm Lines Derived from Triticum turgidum ' T. tauschii Crosses and Resistant to Karnal Bunt. Crop Sci 36:218. doi: 10.2135/cropsci1996.0011183X003600010056x

Walther D, Gindrat D (1988) Biological control of damping-off of sugar-beet and cotton with Chaetomium globosum or a fluorescent Pseudomonas sp. Can J Microbiol 34:631–637. doi: 10.1139/m88-104.

West, J. S., Holdgate, S., Townsend, J. A., Edwards, S. G., Jennings, P., & Fitt, B. D. L. (2012). Impacts of changing climate and agronomic factors on fusarium ear blight of wheat in the UK. Fungal Ecology, 5, 53–61. Doi: 10.1016/j. funeco.2011.03.003.

Wen J-Z, Lu J-Y (1991) Studies on biology of sexual reproduction of Bipolaris sorokiniana. Acta Mycol Sin 10:61–70.

White JW, Rodriguez-Aguilar A (2001) An agroclimatological characterization of Indo-Gangetic Plains. In: Kataki PK (ed) The Rice-Wheat Cropping Systems of South AsiaTrends; Constraints; Productivity and Policy. Food Product Press; New York, pp 53–65.

Wildermuth GB, Thomas GA, Radford BJ, McNamara RB, Kelly A (1997) Crown rot and common root rot in wheat grown under different tillage and stubble treatments in southern Queensland, Australia. Soil Tillage Res 44:211–224.

Yadav B (1981) Behaviour of Cochliobolus sativus during its infection of barley and wheat leaves. Aust J Bot 29:71. doi: 10.1071/BT9810071.

Zhong S, Steffenson BJ (2001) Virulence and Molecular Diversity in Cochliobolus sativus. Phytopathology 91:469–76. doi: 10.1094/PHYTO. 2001.91.5.469.

Zillinsky FJ (1983) Common Diseases of Small Grain Cereals: A Guide to Identification. CIMMYT, Mexico, D.F.

Chapter - 2

An Overview of Markers used for Genetic Diversity Analysis of *Capsicum*

—Md Aminul Islam and Shashi Bhushan Tripathi

2.1 Introduction

The genus *Capsicum* belongs to the family Solancaeae which contains over 40 accepted species. The most common of these are the so called "sweet peppers and "chillies". These species are believed to be originated in South and Central Americas where these were domesticated around 7,000 BC. Christopher Columbus introduced them into Europe during end of 15th century, from where these were spread to rest of the world. *Capsicum* is one of the first crops which were domesticated by man.

Capsicum varieties with pungent fruits are commonly called "hot peppers" or "chillies". Chillies are one of the most consumed spices known for its pungency. When eaten, they give a burning sensation. The word "*Capsicum*" is derived from the Greek word "Kapsimo" meaning "to bite" which is attributed to pungency. In fact, pungency is the trait which made the peppers known to consumers and contributed to its dispersal throughout the world. The heat sensation is due to the presence of a group of alkaloids called capsaicinoids. So far, *Capsicum* is the only genus known to produce capsaicinoids. Pepper fruits are also a good source of vitamins such as A, B, C, E and K. Presence of capsaicinoids has been used as the central trait for the genus *Capsicum* by earlier taxonomists. However, it was found that presence of capsaicinoids is not universal in this genus. The capsaicinoids are synthesized in the epidermal cells of placenta and accumulate there forming blisters. Till date, very little information is available about the genes involved in biosynthesis of capsaicinoids and their regulation.

Genetic markers can be defined as landmarks with known positions in the genome of an organism. During the past two decades, a number of methods have been developed for characterization organisms at DNA level. After the discovery of PCR (polymerase chain reaction), the development and use of molecular markers has increased significantly. Based on PCR, a number of molecular markers have been developed and used in gene tagging and genetic mapping, map based cloning of agronomical important traits, genetic diversity studies, phylogenetic analysis and marker assisted selection (Joshi et al. 2000).

2.2 Molecular marker

Although a large number of molecular marker techniques exist , each marker technique has its own strengths and weaknesses (Powell et al. 1996; Weising et al. 2005). No molecular marker can fit for all requirements. Choice of molecular marker technique is made based on the genome complexity of the species under investigation, the objectives of the study and available infrastructure, expertise and financial support with the researcher.

2.2.1 Morphological markers

Morphological markers are defined as phenotypic characters that can be visualized by naked eyes e.g. size, shape, colour etc. Phenotypic variation is the reflection of variations at the genetic level created due to mutations. Over other genetic markers, morphological markers are fast, easy to detect and economical to use. However, morphological markers are highly influenced by the environment. Other problems related to morphological markers are their ambiguity and limited number.

2.2.2 Biochemical markers

Biochemical markers include structural variants of storage proteins and enzymes with common function but with different molecular weights and electrophoretic mobilities (Tanksley and Orton 1983). Most of these markers are of codominant nature. However, the level of polymorphisms obtained by these markers is very low. This is because only 30 per cent of all nucleotide substitutions in coding regions result in changes in amino acids in the corresponding polypeptide chain due to redundant nature of the genetic code (Lakshmikumaran et al. 1998). These markers were very popular before the advent of DNA based markers. Isozymes of peroxidase and esterase have been used by several workers for studying the genetic variation and cultivar identification in tea (Borthakur et al. 1995; Chen 1996; Singh and Ravindranath 1994; Yang and Sun 1994).

The assay methods of isozyme analysis are tedious and time consuming as compared to modern molecular markers. Consequently, use of these markers is nowadays limited to seed purity tests and occasionally in plant breeding.

2.2.3 DNA based markers

Out of different genetic markers, DNA based markers are the most popular due to their ease of use. A major advantage of DNA based markers over morphological and biochemical markers is that the DNA based markers are not influenced by the environment. A number of molecular markers have been developed over the past two decades. Some of them are described below.

2.2.3.1 Restriction fragment length polymorphism (RFLP)

RFLP is the first molecular marker technique which enabled the detection of polymorphisms in nucleotide sequences of homologous DNA. RFLP was first used by Grodziker et al. (1974) for genetic mapping of temperature sensitive mutation in adeno-virus serotypes. In RFLP, DNA is digested with one or more restriction enzymes which cut the DNA at specific sites. The products of restriction digestion are then separated in agarose gel by electrophoresis and transferred to a membrane. The fragments are then hybridized with labeled probes to detect the specific fragments and variations. RFLP markers have been successfully used in genetic diversity (Keim et al. 1992; Butcher et al. 1998) and linkage mapping in different crops such as rice (Yoshimura et al. 1997), wheat (Chao et al. 1989), maize (Gardiner et al. 1993), barley (Heun et al. 1991) and tomato (Foolad et al. 1998). RFLP analysis requires large amount of highly pure DNA and good supply of probe. RFLP analysis is very laborious, expensive, time consuming with current standards. It is also not amenable to automation.

2.2.3.2 Randomly amplified polymorphic DNA (RAPD)

RAPD was first described by Williams et al. (Williams et al. 1990) and patented by DuPont. It is one of the first molecular marker techniques developed based on PCR. However, unlike conventional PCR, RAPD does not require any specific knowledge of DNA sequences for the target organism. In RAPD, short primers (8-12 nucleotides) with arbitrary sequence are used. Using different combinations of nucleotides, many different RAPD primers can be designed. These markers have been used in a large number of species for genetic diversity analysis and linkage mapping (Eujayl et al. 1997; Ferreira et al. 2000; Oliveira et al. 2004). The major advantages of RAPD are i) requirement of small amounts of DNA (15-25 ng) ii) ease of use iii) and cost effectiveness. However, the reproducibility of this method is low and sensitive to PCR conditions. The RAPD markers are inherited in dominant fashion.

2.2.3.3 Inter simple sequence repeats (ISSR)

ISSR are dominantly inherited markers based on PCR which are similar to RAPD with many aspects. The ISSR technique was first described by (Zietkiewicz et al. 1994). ISSR involves amplification of DNA regions flanked by two identical simple sequence repeats oriented in opposite direction and present within amplification distance. In this technique, the primer sequence (15-30 bp) is composed of microsatellite repeats of different sizes and motifs which have one or two anchored nucleotides at 5' or 3'. This leads to an increase in stringency and reproducibility during amplification as compared to RAPD. ISSR markers have been successfully used in a number of molecular analysis such as varietal identification (Kumar et al. 2001), genetic diversity analysis (Souframanien and Gopalakrishna 2004) and clonal fidelity (Saha et al. 2014; Razaq et al. 2013) in diverse plant species.

2.2.3.4 Simple sequence repeats (SSR)/Microsatellites

Microsatellites, also known as Simple Sequence Repeats (SSRs) or Short Tandem Repeats (STRs) are stretches of DNA that consist of tandem repeats of 1-6 bp. They appear to be ubiquitous in higher organisms, although the frequency of microsatellites varies between species. The term "microsatellite" was coined by Litt and Lutty (1989). The microsatellites are abundant, dispersed throughout the genome and show higher levels of polymorphism than other genetic markers. Their high variability, codominant inheritance and distribution throughout the genome have made them one of the most popular types of markers used in molecular analysis. Microsatellite markers have been the most widely used in the recent years, due to their large number of alleles per locus and their simplistic use by single PCR. However, genotyping of large populations using this technique is still expensive and time consuming.

2.2.3.5 Single nucleotide polymorphism (SNP)

Single Nucleotide Polymorphisms (SNP) and insertions/deletions (INDELs) are the most abundant type of DNA sequence polymorphisms and can be theoretically found within every genomic sequence (Jehan and Lakhanpaul 2006). They can be used as genetic markers for many applications such as cultivar identification, construction of genetic maps, assessment of genetic diversity, detection of genotype/phenotype associations, and marker-assisted selection. Furthermore, the development of high throughput methods for detection of Single Nucleotide Polymorphisms (SNPs) and small indels (insertion/deletions) has led to a revolution in their use as molecular markers. SNPs are increasingly becoming the marker of choice in breeding programs (Gupta et al. 2001). In plants, systematic analysis of nucleotide polymorphism have only been approached in a few, well-studied model species such as Arabidopsis, barley and maize (Rostoks et al. 2005; Zhu et al. 2003; Schmid et al. 2003). Many different techniques can be used to genotype SNP markers, from the low throughput allele specific PCR (Gupta et al. 2008) to high throughput methods genotyping hundreds of thousands of SNP in parallel. Depending on the number of samples and markers to be analyzed, medium to high throughput array based SNP genotyping systems are now available.

2.2.3.6 Amplified fragment length polymorphism (AFLP)

The AFLP technique was first described by Vos et al. (1995) and patented by Keygene N.V. This technique is a combination of RFLP and PCR. This is a highly sensitive method to detect polymorphisms sampled across the genome and has become an important tool in molecular breeding. A flow diagram of the AFLP protocol is shown in figure 2.1. AFLP involves restriction digestion of genomic DNA with two restriction enzymes, a rare cutter (EcoRI/PstI) and, a frequent cutter (MseI/TaqI), with subsequent ligation of synthetic adapters to both the ends of fragments created by restriction digestion. Selective amplification of a set of restriction fragments is usually carried out with labeled primers designed according to the adapter sequence. Amplification products are separated in a high resolving sequencing gel (PAGE) and detected by autoradiography or fluorescence detection. The AFLP technique is much more efficient than other previously developed markers on a number of counts. It is highly reproducible and generates large number of polymorphic markers in a single assay. Like RAPD and ISSR, no prior sequence information is required. AFLP markers are inherited in dominant fashion.

2.2.3.7 Three endonuclease amplified fragment length polymorphism (TE-AFLP)

Van der Wurff et al. (2000) described a new marker technique called TE-AFLP. The basic principles of AFLP and TE-AFLP are identical except the number of restriction enzymes used to generate the library of fragments. Both AFLP and TE-AFLP have been used extensively in this work. A diagrammatic representation of TE-AFLP technique is shown in figure 2.2.

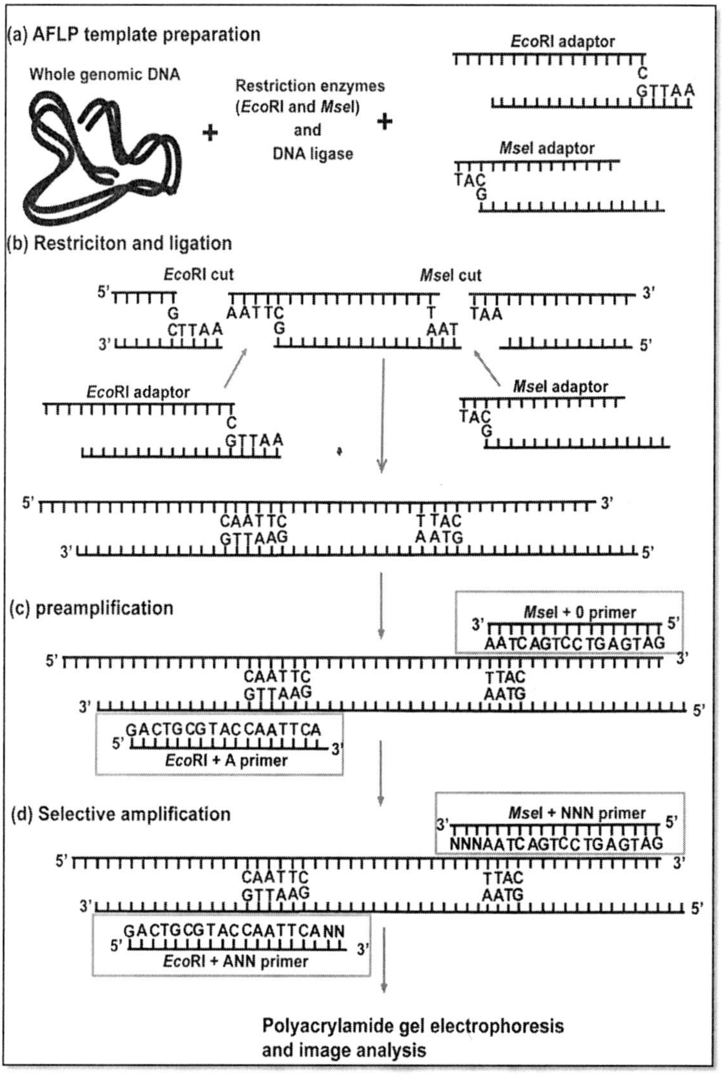

Figure 2.1: A schematic diagram showing different steps of AFLP assay.

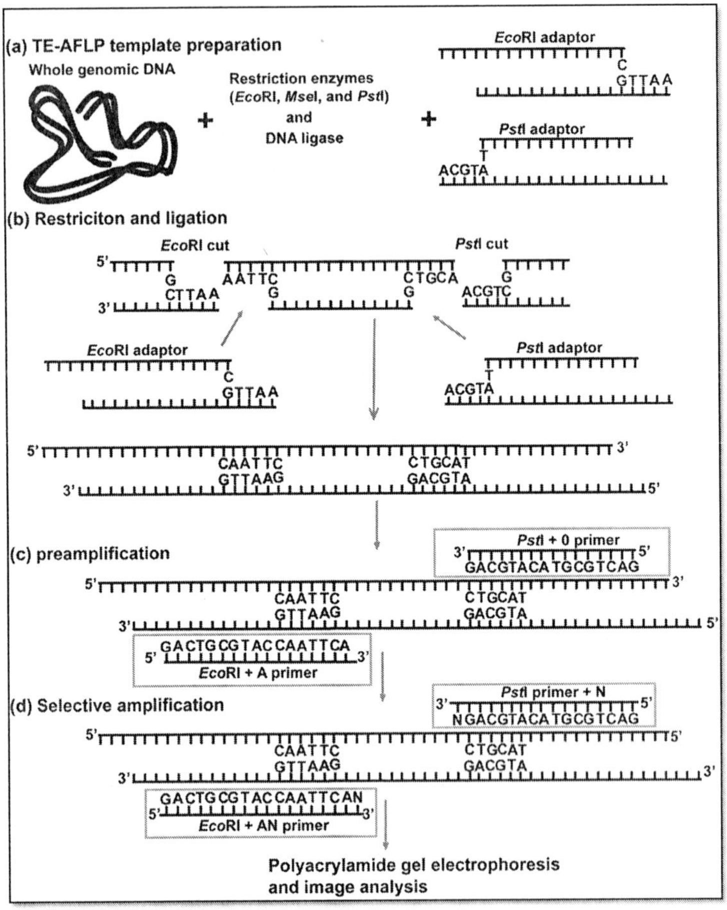

Figure 2.2: A schematic diagram showing different steps of TE-AFLP assay

2.3 Assessment of genetic diversity of Capsicum using molecular markers

Information on genetic diversity in any plant species is an important requirement for undertaking genetic improvement programs (Mohammadi and Prasanna 2003). A large volume of research has been conducted on genetic diversity of *Capsicum* using molecular markers. However, there has been very little research on genetic diversity of *Capsicum* accessions collected from north eastern India. Recently, *Capsicum* accessions from NE India have been characterized using RAPD, SSR, and SNP markers in acyltransferase 3 gene (Yumnam et al. 2012; Sanatombi et al. 2010). However, these studies on NE Indian accessions have used very small sample sizes. There has been no large scale study on genetic diversity of Bhut Jolokia and other landraces of NE India so far.

A number of studies were conducted to characterize the various *Capsicum* germplasm grown across the world. Almost all kind of molecular markers have been used for characterize *Capsicum* germplasm. Some of the important studies of *Capsicum* were briefly discussed in chapter. Loaiza-Figueroa et al. (1989) studied the evolutionary relationships of 186 accessions of *Capsicum* from Mexico using isoenzyme electrophoresis. A total of 76 alleles representing 20 genetic loci coding for nine enzyme systems were observed. The study showed that the *Capsicum* populations in Mexico had predominantly homozygous genotypes presumably due to a self-pollinated breeding system and population bottlenecks. Significant genetic differentiation was found mainly between populations of differing geographical regions. The study suggested that the centre of domestication of *C. annuum* is in Mexico. While, Kumar et al. (2001) employed ISSR and FISSR markers to differentiate four disputed *C. annum* samples. This study distinguished the four samples unambiguously which showed the potential of modern DNA technology to protect the plant breeder rights.

For the first time, Gogoi and Gautam (2002) evaluated genetic variability in 52 genotypes of *Capsicum* representing varieties, breeding lines and local collections from NE India. Wide variations were observed in various morphological characters studied which indicated towards the diverse nature of this germplasm. Similarly, Geleta et al. (2005) evaluated the genetic diversity of representative thirty nine *C. annuum* genotype from different regions of Ethiopia using AFLP and morphological markers. Comparative studies using both the marker systems revealed high genetic diversity in the studied genotypes. Significant and positive correlation between morphological and AFLP markers was obtained which proved the efficacy of both the markers to evaluate the genetic diversity. In the same year, a comparative study was conducted by Tam et al. (2005) to evaluate the usefulness of AFLP, SSR and sequence specific amplified polymorphisms (SSAP) markers for genetic diversity of tomato and *Capsicum*. The study revealed a significant correlation between AFLP, SSR and SSAP markers indicating the efficacy of SSAP markers for diversity analysis. In an another study, Costa et al. (2006) employed RAPD markers to evaluate the genetic diversity of 70 *Capsicum* accessions belonging four different species (*C. annuum, C. baccatum, C. chinense* and *C. frutescens*) and seventeen unspecified genotypes. The study successfully revealed the genetic diversity among genotypes and identified a duplicate accession pair in the germplasm. The study also classified the unknown genotypes to their respective species. Whereas, Thul et al. (2006) used AFLP markers for genetic diversity analysis of 24 accessions of *Capsicum* representing seven species, of which two samples were from the Netherlands and 22 samples were from different geographical locations of India. This study revealed grouping of *Capsicum* accessions based on geographical locations rather than according to their species identity which indicated that a great amount of intermixing of genomic material might have taken in this germplasm. Bhut Jolokia (BJ) is a very highly pungent *Capsicum* landrace grown at northeast India. The taxonomic status of this landrace is unclear. For the first time, Bosland and Baral (2007) used RAPD markers to evaluated the genetic relationship of nineteen genotypes belonging of *C. annuum*, *C. chinense*, *C. frutescence* and Bhut Jolokia. The study identified a number of species specific markers. Bhut Jolokia shared more markers from *C. chinense* than with *C. frutescens*. The average genetic similarity of Bhut Jolokia and *C. chinense* was 0.79 which closer to the mean genetic diversity within *C. chinense* (0.82) than that within *C. frutescens* (0.85). Based on the presence of both *C. frutescens* and *C. chinense* specific markers in Bhut Jolokia, the study concluded it as an interspecific hybrid between the two species. Later on, Bhagowati and Changkija (2009) evaluated the genetic diversity in the *Capsicum* landrace, Naga Jolokia, with morphological characters. The study revealed a considerable amount of genetic variability in this germplasm especially with respect to fruit shape and color.

Similarly, Aktas et al. (2009) used AFLP markers to evaluate the genetic diversity of Turkish pepper varieties. They used 14 Turkish pepper varieties belonging to *C. annuum* along with 5 exotic pepper genotypes. The study revealed a comparatively higher level of genetic diversity in local genotypes than in institutional genotypes. Whereas, Hanacek et al. (2009) used eight SSR markers to evaluate genetic diversity of 41 red peppers (*C. annuum*). A total of 28 alleles were detected, with an average of 3.5 alleles per one microsatellite locus. The study observed low diversity between accessions believed to be distant and genetic distinctness between accessions have same or similar names and suggested that there were possibilities of duplications in the germplasm. Melendez et al. (2009) used three nuclear genes, viz. *Dhn*, *G3pdh* and *Waxy*, to investigate genetic diversity and population structure in semi wild and domesticated landraces of *C. annuum* from Mexico. They observed that the nucleotide diversity of domesticated *C. annuum* genotypes was 10 per cent less as compared to wild *C. annuum* genotypes. The Yucatan Peninsula contained a large number of haplotypes, many of which were unique, suggesting it an important region of chilli domestication and center of diversity.

Ince et al (2009) employed DAMD-PCR markers to identify genotype specific markers. The study involved 15 accessions representing 11 species of *Capsicum* and identified a total of 38 accession specific markers. Later on, Ince et al. (2010) employed Td-RAPD markers to evaluate the genetic diversity of 24 *Capsicum* genotypes representing 11 species. The study revealed four major clusters namely, *C. annuum* complex (CA), *C. baccatum* complex (CB), *C. pubescens* complex (CP), and *C. chacoense* accessions (CA/B) with a mean genetic diversity 0.487. The highest genetic variation was observed among the CP accessions whereas accessions in CB had low level of variations. Whereas, Thul et al.(2009) used data on yield related agronomic traits to evaluate genetic diversity of 24 *Capsicum* genotypes representing six species and different geoclimatic regions of India. This study revealed that taxonomic characters are not always related to the agronomic traits. No association was found between geographical and phenotypic diversity. Fruit diameter, number of fruits per plant, and leaf diameter played the greatest role in differentiation.

Sanatombi et al. (2010) studied the genetic diversity of seven *Capsicum* landraces belonging of *C. annuum*, *C. frutescens* and *C. chinense* grown in Manipur, India, using 10 RAPD markers. Cluster analysis showed two major clusters, where *C. frutescens* and *C. chinense* genotypes grouped in one cluster apart from the *C. annuum* genotypes. The average genetic diversity was more in *C. frutescens* and *C. chinense* genotypes than in *C. annuum*. In the same year, Sudre et al. (2010) carried out genetic diversity of 55 *Capsicum* genotypes using 25 descriptors comprising of 14 morphological and 11 agronomic traits. The study revealed the efficacy of qualitative and agronomic descriptors for species discrimination and identification of genotypes with potential for use in various production sectors. Subsequently, Bhadragoudar and Patil (2011) used RAPD markers to study genetic diversity of 45 Indian genotypes belonging to *C. annuum*. Similarity coefficients in this study ranged between 0.20 to 0.94, which indicated the presence of substantial amount of genetic diversity in this germplasm. Recently, Purkayastha et al. (2012) characterized six genotypes of Bhut Jolokia with eight known *Capsicum* species using three nuclear genes, namely, ITS1, 5.8S and ITS2. The study revealed remarkable genetic differences between Bhut Jolokia, *C. frutescens* and *C. chinense*. However, the study failed to discriminate between *C. frutescens* and *C. chinense*. Simimlarly, Yumnam et al. (2012) evaluated the genetic diversity on 53 chilli landraces of NE India using morphological traits, SSR markers and pun1 locus. They were detected 3-9 alleles for each SSR locus with an average of 5.36 allele/ locus. Cluster analysis based on morphological and SSR markers separated the accessions with campanulate fruits type from those with erect fruit type. This study also identified 79 SNP and 3 indels in pun1 gene showing that this germplasm is a rich source of novel genes. Similar study was conducted by Rai et al. (2013) to evaluate the genetic diversity of *Capsicum* germplasm using SSR and random amplified microsatellite polymorphism (RAMPO) markers. The sample used in this study represented nine countries and included 4 species comprising of germplasm accessions, improved lines and 5 landraces from NE India. The dendrogram based on both the markers showed two major clusters and was able to separate *C. annuum* genotypes from others. The study revealed the presence of higher genetic diversity in other *Capsicum* spp. than *C. annuum* genotypes. Recently, Yatung et al. (2014) evaluated genetic diversity of Indian *C. annuum* genotypes using biochemical and morphological markers. Cluster analysis placed the 30 genotyped into 6 groups. The results revealed that capsaicin content and ascorbic acid had the maximum contribution towards genetic divergence.

2.4 Conclusion

Capsicum is one of the earliest crops domesticated by humans. Major breeding efforts have been to produce the commercially attractive varieties for human consumption as vegetable and spices and utilization of capsaicinoids in pharmaceutical applications. However, due to over selection, the genetic base of cultivated *Capsicum* is gradually decrease. Wild relatives of a crop species are known to be important reservoirs of phenotypic and genetic diversity harbouring many alleles of agronomic importance. Not exceptionally wild relatives of *Capsicum* exhibited more abiotic and biotic stress tolerance compared to cultivars. The identification of genetic determinants underlying such variation will aid crop improvement. Thus, identified alleles can be employed for development of superior genotypes. However, the exploitation of such superior allele is dependent on genome compatibility between cultivated and wild relatives, and methods for horizontal gene transfer. Therefore, to considering the importance of wild relatives of *Capsicum*, we must increase the germplasm for future crop improvement program.

References

Aktas H, Abak K, Sensoy S (2009) Genetic diversity in some Turkish pepper (*Capsicum annuum* L.) genotypes revealed by AFLP analyses. Afr J Biotechnol 8:4378-4386.

Bhadragoudar MR, Patil CG (2011) Assessment of genetic diversity among *Capsicum annuum* L. genotypes using RAPD markers. Afr J Biotechnol 10:17477-17483.

Bhagowati RR, Changkija S (2009) Genetic variability and traditional practice in Naga king chili landrace of Nagaland. Asian Agrihist 13:171-180.

Borthakur S, Mondal TK, Borthakur A, Deka PC (1995) Variation in peroxidase and esterase isoenzymes in tea leaves. Two a Bud 42:20-23.

Bosland PW, Baral JB (2007) "Bhut Jolokia"-The world's hottest known chile pepper is a putative naturally occurring interspecific hybrid. HortScience 42:222-224.

Butcher PA, Moran GF, Perkins HD (1998) RFLP diversity in the nuclear genome of *Acacia mangium*. Heredity 81:205-213.

Chao S, Sharp PJ, Worland AJ, Warham EJ, Koebner RMD, Gale MD (1989) RFLP-based genetic maps of wheat homologous group 7 chromosomes. Theoretical and Applied Genetics 78:495-504.

Chen C (1996) Analysis on the isozymes of tea plants F_1 hybrids. Journal of Tea Science 16:31-33.

Costa FRd, Pereira TNS, Vitória AP, Campos KPd, Rodrigues R, Silva DHd et al. (2006) Genetic diversity among *Capsicum* accessions using RAPD markers. Crop Breed Appl Biotechnol 6:18-23.

Eujayl I, Baum M, Erskine W, Pehu E, Muehlbauer FJ (1997) The use of Rapd markers for lentil genetic mapping and the evaluation of distorted F_2 segregation. Euphytica 96:405-412.

Ferreira AR, Foutz KR, Keim P (2000) Soybean genetic map of RAPD markers assigned to an existing scaffold RFLP map. Journal of Heredity 91:392-396.

Foolad MR, Chen FQ, Lin GY (1998) RFLP mapping of QTLs conferring salt tolerance during germination in an interspecific cross of tomato. Theory of Applied Genetics 97:1133-1144.

Gardiner JM, Coe EH, Hancock SM, Hoisington DA, Chao S (1993) Development of a core RFLP map in maize using an immortalized F2 population. Genetics 134:917-930.

Geleta LF, Labuschagne MT, Viljoen CD (2005) Genetic variability in pepper (*Capsicum annuum* L.) estimated by morphological data and amplified fragment length polymorphism markers. Biodiversity and Conservation 14:2361-2375.

Gogoi D, Gautam BP (2002) Variability, heritability and genetic advance in chilli *(Capsicum spp.)*. Agric Sci Dig 22:102-104.

Grodzicker TJ, Williams PS, Sambrook J (1974) Physical mapping of temperature-sensitive mutations of adenovirus. Cold Spring Harbor Symp Quant Bio 39:439-445.

Gupta PK, Roy JK, Prasad M (2001) Single nucleotide polymorphisms: a new paradigm for molecular marker technology and DNA polymorphism detection with emphasis on their use in plants. Current science 80:524-535

Gupta PK, Rustgi S, Mir RR (2008) Array-based high-throughput DNA markers for crop improvement. Heredity 101:5-18.

Hanacek P, Vyhnanek T, Rohrer M, Cieslarova J, Stavelíkova H (2009) DNA polymorphism in genetic resources of red pepper using microsatellite markers. HortScience(Prague) 36:127-132.

Heun M, Kennedy AE, Anderson JA, Lapitan NLV, Sorrells ME, Tanksley SD (1991) Construction of a restriction fragment length polymorphism map for barley (*Hordeum vulgare*). Genome 34:437-447.

Ince AG, Karaca M, Onus AN (2009) Development and utilization of diagnostic DAMD-PCR markers for *Capsicum* accessions. genetic Resources and Crop Evolution 56:211-221.

Ince AG, Karaca M, Onus AN (2010) Genetic relationships within and between *Capsicum* species. Biochemical Genetics 48:83-95.

Jehan T, Lakhanpaul S (2006) Single nucleotide polymorphism (SNP) methods and applications in plant genetics: a review. Indian Journal of Biotechnology 5:435-459.

Joshi SP, Gupta VS, Aggarwal RK, Ranjekar PK, Brar DS (2000) Genetic diversity and phylogenetic relationship as revealed by inter simple sequence repeat (ISSR) polymorphism in the genus Oryza. Theoretical and Applied Genetics 100:1311-1320.

Keim P, Beavis W, Schupp J, Freestone R (1992) Evaluation of of soybean RFLP marker diversity in adpated germplasm. Theory of Applied Genetics 85: 205-212.

Kumar LD, Kathirvel M, Rao GV, Nagaraju J (2001) DNA profiling of disputed chilli samples (*Capsicum annum*) using ISSR-PCR and FISSR-PCR marker assays. Forensic Science International 116:63-68.

Lakshmikumaran M, Das S, Rajagopal J, Goswami J, Negi MS, Bhatia S. 1998. Repeated DNA sequences in Plants: Organization, Evolution and Applications. In: Gupta, P. K., Singh, S. P., Balyan, H. S., Sharma, P. C. and Ramesh, B., editors. Genetics and biotechnology in crop improvement. Meerut, India: Rastogi publications. p. 63-93.

Litt M, Luty JA (1989) A hypervariable microsatellite revealed by in-vitro amplification of a di-nucleotide repeat within the cardiac muscle action gene. American Journal of Human Genetics 44:397-401.

Loaiza-Figueroa F, Ritland K, Cancino JAL, Tanksley SD (1989) Patterns of genetic variation of the genus *Capsicum* (Solanaceae) in Mexico. Plant Systematics and Evolution 165:159-188.

Melendez AA, Morrell Pl, Roose Ml, kim SC (2009) Genetic diversity and structure in semiwild and domesticated chiles (*Capsicum annuum*; solanaceae) from Mexico. American Journal of Botany 96:1190-1202.

Mohammadi SA, Prasanna BM (2003) Analysis of genetic diversity in crop plants: Salient statistical tools and considerations. Crop Science 43:1235-1248.

Oliveira RPd, Vildoso CIA, Cristofani M, Machado MA (2004) Skewed RAPD markers in linkage maps of Citrus. Genetics and Molecular Biology 27: 437-441.

Powell W, Morgante M, Andre C, Hanafey M, Vogel J, Tingey S et al. (1996) The comparison of RFLP, RAPD, AFLP and SSR (microsatellite) markers for germplasm analysis Molecular Breeding 2:225-238.

Purkayastha J, Alam SI, Gogoi Hk, Singh L, Veer V (2012) Molecular characterization of 'Bhut Jolokia' the hottest chilli. Journal of Biosciences 37:757-768.

Rai VP, Kumar R, Kumar S, Rai A, Kumar S, Singh M et al. (2013) Genetic diversity in *Capsicum* germplasm based on microsatellite and random amplified microsatellite polymorphism markers. Physiol Mol Bio Plants 19:575-586.

Razaq M, Heikrujam M, Chetri SK, Agrawal V (2013) In vitro clonal propagation and genetic fidelity of the regenerants of *Spilanthes calva* DC. using RAPD and ISSR marker. Physiol Mol Bio Plants 19:

Rostoks N, Mudie S, Cardle L, Russell J, Ramsay L, Booth A et al. (2005) Genome-wide SNP discovery and linkage analysis in barley based on genes responsive to abiotic stress. Molecular Genetics and Genomics 274: 515-527.

Saha S, Sengupta C, Ghosh P (2014) Evaluation of the genetic fidelity of in vitro propagated *Ocimum basilicum* L. using RAPD and ISSR markers. Journal of Crop Science and Biotechnology 17:281-287.

Sanatombi K, Sen-Mandi S, Sharma GJ (2010) DNA profiling of *Capsicum* landraces of Manipur. Sci Hort 124:405-408.

Schmid KJ, Sorensen TR, Stracke R, Torjek O, Altmann T, Mitchell-Olds T et al. (2003) Large-scale identification and analysis of genome-wide single-nucleotide polymorphisms for mapping in *Arabidopsis thaliana*. Genome Research 13:1250-1257.

Singh HP, Ravindranath SD (1994) Occurrence and distribution of PPO Activity in floral organs of some standard and local cultivars of tea. Journal of the Science of Food and Agriculture 64:117-120.

Souframanien J, Gopalakrishna TG (2004) A comparative analysis of genetic diversity in black gram genotypes using RAPD and ISSR markers. Theoretical and Applied Genetics 109:1687-1693.

Sudre CP, Goncalves LSA, Rodrigues R, Junior ATdA, Riva-Souza EM, Bento CdS (2010) Genetic variability in domesticated *Capsicum* spp as assessed by morphological and agronomic data in mixed statistical analysis. Genetics and Molecular Research 9:283-294.

Tam SM, Mhiri C, Vogelaar A, Kerkveld M, Pearce SR, Grandbastien MA (2005) Comparative analyses of genetic diversities within tomato and pepper collections detected by retrotransposon-based SSAP, AFLP and SSR. Theoretical and Applied Genetics 110:819-831.

Tanksley SD, Orton TJ. 1983. Isozymes in plant genetics and breeding. Amsterdam, The Netherlands: Elsevier Science Publishers.

Thul ST, Lal RK, Shasany AK, Darokar MP, Gupta AK, Gupta MM et al. (2009) Estimation of phenotypic divergence in a collection of *Capsicum* species for yield-related traits. Euphytica 168:189-196.

Thul ST, Shasany AK, Darokar MP, Khanuja SPS (2006) AFLP analysis for genetic diversity in Capsicum annuum and related species. Nat Prod Commun 1:223-228.

van der Wurff AWG, Chan YL, van Straalen NM, Schouten J (2000) TE-AFLP: combining rapidity and robustness in DNA fingerprinting. Nucleic Acids Research 28:e105.

Vos P, Hogers R, Bleeker M, Reijans M, van de Lee T, Hornes M et al. (1995) AFLP: a new technique for DNA fingerprinting. Nucleic Acids Research 23:4407-4414.

Weising K, Nybom H, Pfenninger M, Wolff K, Kahl G 2005. DNA Fingerprinting in Plants: Principles, Methods, and Applications. CRC Press, Taylor and Francis Group, 6000 Broken Sound Parkway NW Boca Raton.

Williams JG, Kubelik AR, Livak KJ, Rafalski JA, Tingey SV (1990) DNA polymorphisms amplified by arbitrary primers are useful as genetic markers. Nucleic Acids Research 18:6531-6535.

Yang Y, Sun T (1994) Study on the esterase isoenzyme in tea mutagenic breeding. China Tea 16:4-9.

Yatung T, Dubey RK, Singh V, Upadhyay G (2014) Genetic diversity of chilli (Capsicum annuum L.) genotypes of India based on morpho-chemical traits. Aust J Crop Sci 8:97-102.

Yoshimura A, Ideta O, Iwata N (1997) Linkage map of phenotype and RFLP markers in rice. Plant Molecular Biology 35:49-60.

Yumnam JS, Tyagi W, Pandey A, Meetei NT, Rai M (2012) Evaluation of genetic diversity of chilli landraces from North Eastern India based on morphology, SSR markers and the Pun1 locus. Plant Molecular Biology Reporter 30:1470-1479.

Zhu YL, Song QJ, Hyten DL, Van TCP, Matukumalli LK, Grimm DR et al. (2003) Single-nucleotide polymorphisms in soybean. Genetics 163:1123-1134.

Zietkiewicz E, Rafalski A, Labuda D (1994) Genome Fingerprinting by Simple Sequence Repeat (SSR)-Anchored Polymerase Chain Reaction Amplification. Genomics 20:176-183.

Chapter - 3

Engineering Pod Shatter Resistance in Crop Brassicas using Artificial miRNA Technology

— *Priyanka Dhakate, S.M. Shivaraj, Anandita Singh*

3.1 Introduction

Rapeseed-mustard oil ranks third in world oilseed production after soybean (*Glycine max*) and palm (*Elaeis guineensis*) oil (Chauhan et al. 2006; Shekhawat et al. 2012). Contribution to global rapeseed production from India stands third at around 6.7 Mt after European Union (10–13 Mt) and China (11-12 Mt) (Shekhawat et al. 2012). In world acreage and production, India contributes 28.3 per cent and 19.8 per cent, respectively (Shekhawat et al. 2012; Bandopadhyay et al. 2013). Domestically, rapeseed mustard is second to groundnut production and accounts for nearly one-third of the oil produced (Shekhawat et al. 2012). Rapeseed cultivation in India chiefly includes Indian mustard, sarson (yellow and brown) and toria crops. Of these crops, Indian mustard (*Brassica juncea* (L.) is chiefly cultivated in Uttar Pradesh, Haryana, Rajasthan, Gujarat and Madhya Pradesh. In southern part of India, it is cultivated in some areas of Andhra Pradesh, Tamil Nadu and Karnataka (Meena et al. 2014; Shekhawat et al. 2012).

Technological interventions applied primarily through conventional breeding approaches led to a so called "yellow revolution" in the country a few decades back (Dastagiri et al. 2014; Shekhawat et al. 2012). A significant increase in productivity and production from 650 kg/ha and 2.68 Mt in 1985-86 to 1022 kg/ha and 6.96 Mt in 1996-1997, respectively, were reported (Gupta and Lal 2011; Dastagiri et al. 2014). In the years that followed, mustard cultivation in India was anticipated to set new benchmarks, however, distinct gaps appeared between production potential and actual realized potential of the mustard crop. For instance, the average productivity in India currently stands at only 791 kg ha^{-1} compared to the world average of 1718 kg ha^{-1} (Gill et al. 2012). In recent years, although the area of mustard cultivation increased, the crop productivity has dropped markedly owing to the impact of various environmental stresses on growth and development of plants (Gill et al. 2012). Hence, mitigating yield losses and simultaneously augmenting crop productivity under unfavorable conditions is a challenge for modern day agriculture (Gill et al. 2012). Particularly in context to climate change, there is an urgent need to focus on reshaping science and technological interventions to boost crop production. In India, mustard breeding programs are concentrated largely on improving oil quality. However, only limited efforts are being made in developing varieties that are resistant to abiotic and biotic stresses. Complex mechanisms of abiotic stress tolerance regulated by coordinated gene expression of multiple genes have impeded the rate and extent of progress in improving abiotic and biotic stress tolerance in crops (Gill et al. 2012). In such a scenario, developing cultivars with stress evading strategies promise mitigation of yield losses despite continued onslaughts of environmental stresses. Adaptive strategies such as synchronized seed dispersal promises better crop response to added inputs. Synchronized seed dispersal ensures reduction in yield losses at the time of harvesting (Gill et al. 2012). Biotechnological methods offer viable options to develop such plant varieties with ease and efficiency. Therefore, transgenic technology is better poised for swift generation of novel varieties expressing selective traits, although this technology relies on availability of well characterized candidate genes.

To achieve this here we present application of a silencing technology involving design of artificial miRNA and its deployment to redundantly silence homologs of *SHATTERPROOF1* (*SHP1*) and *SHATTERPROOF2* (*SHP2*) from *B. napus*. In Canada and Europe, *B. napus* (canola) is the most extensively grown oilseed crop. Globally, canola oil is more preferred over rapeseed mustard due to low erucic acid and glucosinolate (Priyamedha et al. 2015; Javed et al. 2014). However, this crop faces major yield losses of up to 50% on field due to asynchronous shattering of pods (Macleod 1981). In view of lack of an efficient alternative strategy for curbing such yield losses, we present a proof of concept of deploying amiRNA technology to achieve shatter resistance in canola. Such target specific redundant silencing of *SHATTERPROOF* (*SHP1*) and *SHATTERPROOF2* (*SHP2*) transcripts that control valve margin formation for fruit dehiscence in *Arabidopsis* (Liljegren et al. 2000) and Brassicas (Squires et al. 2003) shall bring about pod shatter resistance in canola crops. This technology is beneficial for use in polyploid genomes of Brassicas such as *B. napus*, wherein simultaneous silencing of all *SHP1*/*SHP2* transcripts is required to confer shatter resistance.

3.2 Fruit Dehiscence: Molecular Mechanism

3.2.1 Fruit morphology and dehiscence mechanism in Brassicas

The fruit in plants primarily provides for nourishment and security of the developing seeds before they are dispersed. At an appropriate time, the pod shatters releasing the mature seed for germination. Fruit dehiscence as an adaptive strategy ensures efficient seed dispersal thus contributing to evolutionary success of the plants. Synchronous growth of the developing fruits and seeds is a result of a well-orchestrated differentiation and regulation of numerous cell types

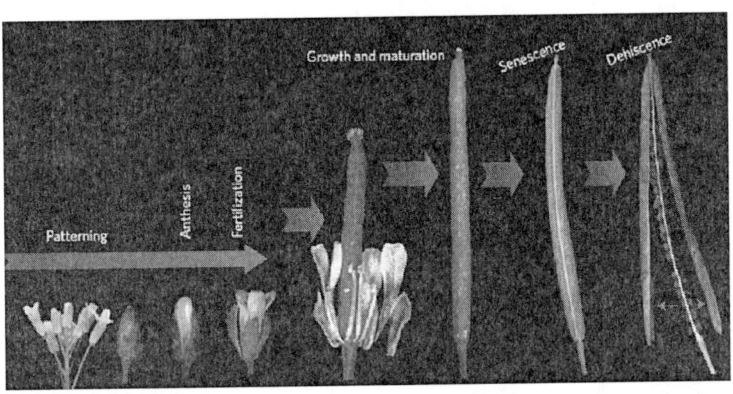

Figure 3.1: Stages of fruit dehiscence in A. thaliana. [Image credit: (2015)].

present in this inconspicuous yet complex plant organ. Extensive genetic studies in *A. thaliana* have led to basic understanding of the molecular mechanisms that regulate fruit development and dehiscence (Giovannoni 2004; Seymour et al. 2008; Seymour et al. 2013). Through comparative physiological, anatomical and molecular studies similar process of have also been revealed in *B. rapa and B. napus* (Spence et al. 1996) as well as many crop species such as tomato, canola, and rice. These study models aid in dissection of similar processes that modulate seed dispersal.

Fruit development is primarily divided into distinct phases. The pre fertilization phase relates to specification and development of the gynoecium while post fertilization phase explains fruit dehiscence (Figure 3.1). In *A. thaliana,* after fertilization, the carpel transforms into a silique which emerges from a bilocular ovary with synchronized development of ovules that later mature into seeds. In a typical silique, three distinct structures are present: the replum, the valves and the valve margin (Figure 3.2a). In later stages of silique development, a conspicuous elongation of the valves occurs followed closely by maturation and a dramatic fruit dehiscence phenomenon commonly

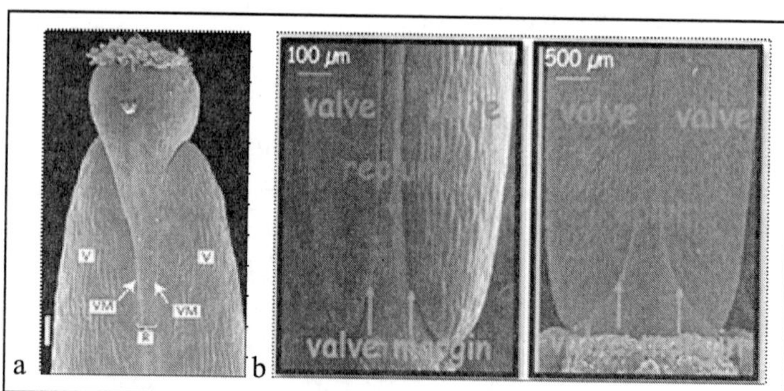

Figure 3.2 a. A mature A. thaliana silique. The two (V) derived from the ovary walls are connected with each other by the replum (R). During fruit dehiscence, the valve margins (VM), define the future dehiscence zone that mediates separation of valves from the replum. (Image credit Ripoll et al. 2015). b. Mature siliques from A. thaliana (right) and Brassicas (left) share same anatomy. b. The mature siliques from A. thaliana and share similar anatomy (Image credit: Ostergaard et al. 2007).

known as 'pod shatter'. A. thaliana and Brassicas share a common silique anatomy method of seed dispersal (Figure 3.2b). In A. thaliana, such dramatic seed dispersal mechanisms ensures effective seed dispersal (Nathan and Muller-Landau 2000; Dong and Wang 2015). However, in Brassica crop plants, such asynchronous 'pod shattering' translates into huge losses in seed yields.

3.2.2 Key regulators determining pod shattering in Brassicas

Regulatory networks controlling fruit patterning, differentiation and dehiscence are well understood in A. thaliana. At the time of seed dispersal, the two valves from the septum are detached (Bowman et al. 1999; Ferrándiz et al. 1999; Pesaresi et al. 2014) via formation of a zone of dehiscence (DZ) which develops as valve margins of the silique (Figure 3.2). In this zone, a few cells divide and extend throughout the entire length of the fruit. The first set of genes that were characterized and implicated in formation of DZ were MADS box transcription factors, SHATTERPROOF1 (SHP1) and SHATTERPROOF2 (SHP2) (Liljegren et al. 2000; Dardick and Callahan 2014; Pesaresi et al. 2014; Mühlhausen et al. 2013). These genes are redundantly required for proper specification of DZ. Single mutants of these genes showed no phenotype. In the double mutant (shp1/shp2), however, the siliques are unable to shatter post fruit maturity implying redundant functions (Liljegren et al. 2000; Liljegren et al. 2004; Dardick and Callahan 2014; Pesaresi et al. 2014; Mühlhausen et al. 2013). SHP1 and SHP2 are responsible for promoting lignification of the adjacent cells in DZ to bring about shattering of the pods. Another gene, INDEHISCENT, is expressed in the valve margins (Liljegren et al. 2004). On the other hand, ALCATRAZ (ALC) governs the development of the separation layers (Rajani and Sundaresan 2001) SHP1 and SHP2 promote the expression of INDEHISCENT and ALC, thus facilitating lignification of DZ cells (Liljegren et al. 2000). Another MADS box gene, FRUITFULL (FUL) is known to regulate DZ formation. FUL expression is complementary to SHP1/SHP2 and is restricted to valve margin cells (Gu et al. 1998; Mandel and Yanofsky 1995; Chung et al. 2013; Mühlhausen et al. 2013; Østergaard 2009). SEEDSTICK (STK) MADS box gene also acts redundantly with SHP1 and SHP2 to repress FUL expression in ovules and seeds (Favaro et al. 2003; Pinyopich et al. 2003; Chung et al. 2013; Mühlhausen et al. 2013). Additionally, INDEHISCENT1 (IND1) is also required for DZ formation, as ind1 mutant siliques were also unable to shatter (Liljegren et al. 2004; Girin et al. 2010). ALCATRAZ (ALC) is known to control valve margin formation (Rajani and Sundaresan 2001; Mühlhausen et al. 2013). Interplay of these genes is responsible for seed dispersal in Arabidopsis. Of these, SHP1/ and SHP2, are at the top of signaling cascade for regulation of pod shattering in Arabidopsis. SHP orthologs have been reported from plant species including; Vitis vinifera (Boss et al. 2001); Malus domestica (Van der Linden et al. 2002), Prunus persica (Tani et al. 2007).

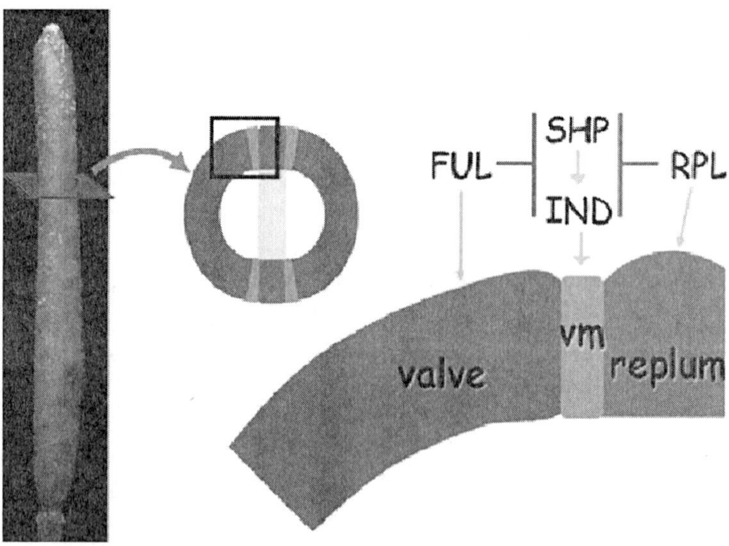

Figure 3.3:* *Diagrammatic representation of molecular mechanisms determing pod shattering in A. thaliana

Taihangia rupestris (Lü et al. 2007), *Phaseolus vulgaris* (Nanni et al. 2011), *Solanum esculentum* (Vrebalov et al. 2009), *Citrus sinensis*

(Araújo et al. 2013), and *Brassica rapa* and *B. napus* (Spence et al. 1996). Even though fruit dehiscence is well understood in Brassicaceae, manipulation of this trait for curbing yield losses through biotechnological approaches is still at its nascent stages. My study embodies a strategy for manipulation of fruit dehiscence in Brassicas

3.2.3 Approaches to raise pod shatter resistant canola crops.

Our understanding of the mechanistics aspects of seed shatter processes in Brassicas and other crops has seen significant advancement over the past decade. With increase in conceptual knowledge of these processes, strategies to manipulate shatter are also in a dynamic process of development and evolution. These crop improvement programs, therefore, aim at minimizing grain loss in the mature crop during harvesting.

Raising a pod shatter canola crop canola (*Brassica napus* L.) is one of the major target traits for global rapeseed production (Kuai et al. 2015; Peng et al. 2015; Østergaard 2009). One of the first attempts to translate available genetic information for introgression of pod shatter resistance in *Brassica juncea* was carried out by over-expression *AtFUL* to make pods resistant to shattering (Østergaard 2009). Globally, canola oil is more preferred over rapeseed mustard due to low erucic acid and glucosinolates content. However, this crop faces major yield losses of up to 50 per cent on field due to asynchronous shattering of pods (Macleod 1981). For curbing such yield losses, conventional breeding approaches are underway to transfer shatter resistance genes from shatter tolerant cultivars *B. juncea* and *B. rapa* to *B. napus* by conventional breeding methods. Meanwhile, Bayer CropScience has recently launched a shatter tolerant canola variety, InVigor L140P that is being widely grown across Canada since 2012 (Bayer Crop Science Canada). Other strategies such as pod sealants Pod Ceal DC (http:// www.brettyoung.ca /Western_Canada /Products/ Pod_Ceal/podcea l_products) and Pod-Stik (United AgriProducts 2011) available to farmers to reduce the risk of pod shatter as the seeds mature. These "sealants" are sprayed across the crop when approx 30-40 per cent of the pods are pliable, not brittle and have changed color. However, yield benefits from this technology remain debatable (Holzapfel et al. 2010).

Recently, several quantitative trait loci (QTL) controlling this trait were identified unraveling the genetic basis of natural variation for shatter resistance in *B. napus* (Raman et al. 2014). These findings provide a plethora of information to the breeders for targeting introduction of useful alleles from other related genotypes of *Brassica*. Yet, developing synchronous pod shatter and (or) pod shatter resistance in canola is still challenging. Hormonal treatments have been studied to induce synchronous pod shatter regularly (Child et al. 1999; Kuai et al. 2015). While reduction in ethylene levels in the pods delays maturation it does not affect pod shatter. However, sustained auxin application in maturing pods does seem to induce pod shatter resistance (Bennett et al. 2011; Dong and Wang 2015). Crop breeding programs using genetic manipulation of the hormonal and genetic mechanisms have been by far very encouraging. A strong correlation between ectopic expression of an endo-acting isoform of polygalacturonase (PG35-8) and the breakdown of the middle lamella of DZ cells was observed in canola crops (Jaradat et al. 2014).

In view of lack of an efficient alternative strategy to achieve shatter resistance in canola, *I* present a proof of concept of deploying amiRNA technology to curb yield losses in Brassicas. *SHATTERPROOF* (*SHP1*) and *SHATTERPROOF2* (*SHP2*) are MADS-box genes that redundantly control formation of valve margin for fruit dehiscence in *Arabidopsis* (Liljegren et al. 2000) and also in oil-seed rape (Squires et al. 2003). At the time of fruit dehiscence, these genes promote lignification of adjacent cells in the zone of dehiscence, thereby, resulting in shattering of the pod (Liljegren et al. 2000). Hence, in complex genomes of Brassicas such as *B. napus*, simultaneous silencing of all *SHP1*/*SHP2* transcripts is anticipated to confer shatter resistance.

3.3 The amiRNA Technology

RNAi mechanisms have been implicated for efficient manipulation desired traits in crops over last decade (Ambesajir et al. 2012; Duan et al. 2012; Younis et al. 2014). These mechanisms have typically used principles rooted in inherent post transcriptional gene silencing (PTGS) mechanisms of plants to bring about target specific silencing. These strategies are classically relying on the production of small interfering RNAs (siRNAs) that are derived from dsRNAs and introduced to the plant genome. One such is known as virus-induced gene silencing (VIGS) that uses viral genomes as vectors to introduce dsRNAs to target and silence a desired plant genes (Lu et al. 2003; Sablok et al. 2011). Although this approach results in target specific silencing, identification of an appropriate viral vector with causing pathological symptoms in the plant is a major challenge (Unver and Budak 2009). Alternative strategies employ production of siRNAs by over-expression of the target gene in antisense direction (Jorgensen et al. 2006) that result in sense-antisense dsDNA fragments that may or may not include a hairpin (Helliwell and Waterhouse 2005). However, given the fact that siRNA signal is amplifiable due to production of numerous siRNAs from the long dsRNA fragments introduced, this strategy carries an inherent risk of silencing undesired genes (off- target silencing) (Schwab et al. 2006; Xu et al. 2006; Sablok et al. 2011). In addition, this approach can also lead to auto silencing of the target gene (Molnar et al. 2009). To circumvent these challenges, extensive research in understanding miRNA based silencing in plants led to establishment of the amiRNA technology for target based gene silencing.

To design amiRNAs, the targets are assayed to identify potential amiRNA binding sites (20 bp) (Dhakate et al. 2014). The amiRNAs are then engineered into an endogenous miRNA precursor wherein the miRNA:miRNA* region is replaced with amiRNA:amiRNA* region. The efficiency of the new precursor to process amiRNA remains unaltered until its secondary fold back structure is intact. Therefore, mature amiRNA are efficiently processed in the plant genome and brings about target specific silencing of the desired gene (Schwab et al. 2006; Schwab et al. 2010; Sablok et al. 2011; Dhakate et al. 2014). Since amiRNAs are designed to direct silencing of targets that harbor corresponding binding sites, they can be employed for functional characterization of gene(s) while avoiding the risk of off-target silencing(Parizotto et al. 2004; Park et al. 2005; Sablok et al. 2011). In addition amiRNAs are found stable through various generations (Park et al. 2009; Zhao et al. 2009; Molnar et al. 2009). Further, since amiRNAs are small in size it is possible to engineer multiple and unrelated amiRNAs in the same construct that can target multiple genes, alleles or splice forms in a target specific manner at the same time (Lin et al. 2009, Niu et al. 2006, Schwab et al. 2010). Further, in comparison to other RNAi technologies amiRNAs have fewer biosafety concerns when applied to agriculture on a large (Duan et al. 2008; Liu and Chen 2010). amiRNAs have been used increasingly to understand underlying mechanisms in Arabidopsis (Parizotto et al. 2004; Kurihara and Watanabe 2004) such as auxin signaling (Michniewicz et al.), circadian rhythms (Maier et al. 2009; Kim and Somers 2010), pollen development (Pereira et al. 2006; Levitin et al. 2008). Use of amiRNAs for functional characterization of genes and crop trait manipulation has been well summarized by Sablok et al. (2011).

Given the efficiency of amiRNA to specifically target multiple genes at the same time in the present study, I propose to achieve redundant silencing of *Brassica SHP1* and *SHP2* homologs for preventing unintended pod shatter. In a polyploid species such as *B. napus*, most genes are present in multiple copies, hence silencing of all native *SHP1/SHP2* transcripts becomes a major challenge. To this effect, a highly specific gene silencing approach, such as amiRNA technology is more preferred than siRNAs Principles governing natural miRNA-based target selection are used by graphical interfaces such as Web MicroRNA Designer 3 (WMD3) to predict candidate amiRNAs (http://wmd3.weigelworld.org/). To compute optimal amiRNA sequences, WMD tool uses target genes as an input sequence and generates a list of all specific amiRNAs with the top candidate predicted with highest possibility of directing target cleavage (Schwab et al. 2006). amiRNAs resemble natural miRNAs primarily on three parameters: the sequences start with U, they display 5' instability relative to the amiRNA*, and the 10th nucleotide is either an A (Reynolds et al. 2004) or an U. Like mentioned previously, so long as the secondary fold back structure of the precursor remains unaltered, the efficiency of the designed precursor (pre-amiRNAs) remains stable as the native precursor. Therefore, for *Arabidopsis* (or other dicot species) MIR319a precursor backbone is used to engineer the predicted amiRNA sequence by replacing the original miR319a sequence with the amiRNA sequence. When the new precursor backbone is introgressed into plant genome, the mature amiRNA sequence is expected to be efficiently processed to direct the cleavage of desired target(s). However, after amiRNA is predicted using wmd3 tool, *in silico* predictions can be used to help screen most efficient candidates for *in-vivo* validation demonstration of cleavage of the intended targets (Dhakate et al. 2014).

Stability of secondary structure of precursor amiRNAs governing amiRNA biogenesis can be evaluated using RNAfold web server (http://rna.tbi.univie.ac.at/cgi-bin/RNAfold.cgi). Furthermore, the total free energy of binding (TFEB) that defines the interaction between 80-bp sequence contexts of target site within transcripts with amiRNAs is evaluated using RNAup. TFEB is calculated as sum total of energy gained from duplex formation (EDF), energy lost as opening energy for the longer sequence (OELS) and opening energy for the shorter sequence (OESS). Suitable amiRNAs after this primary screening can be used in *N. benthamiana* based transient assays to validate both amiRNA biogenesis and their efficiency of directing cleavage of intended targets (Dhakate et al. 2014). Competent amiRNAs candidates post this initial screening can be used to introgressed in desired plant genome for either functional characterization of genes and (or) crop trait manipulation

Artificial miRNA based technology based reverse genetic approaches are thus more favorable in polyploid genomes where loss-of-function mutant backgrounds are hard to obtain owing to gene redundancy. Yet the only challenge of employing of amiRNA technology in polyploid genomes lies in designing a unique amiRNA that can simultaneously and specifically silence redundant gene transcripts. Hence, an assessment of the target transcripts *a priori* is an essential pre-requisite for design of amiRNA.

3.4 Conclusion

In rapidly changing climate and environment, developing resilient crops though biotechnological approaches promise a viable solution for modern day sustainable agriculture. Crops armoured with stress evading abilities are similarly relevant. A measure to mitigate yield losses stems from synchronizing seed dispersal in crop plants. Evolutionary forces have enabled natural selection of plants with elaborate survival strategies. One of these adaptive strategies is natural seed dispersing mechanisms that ensure reproductive success of the plants across varied habitats. However, crops are selected against natural seed dispersal as they result in tremendous yield losses In domestication of seed crops such as cereals, legumes and oilseeds, non-dehiscent trait is considered as a milestone as it renders a crop dependent' on human interventions for its propagation and further facilitates fixation of desired characters Similarly, while in *A. thaliana*, spontaneous pod shatter ensures efficient seed, in Brassica crops such dramatic pod shatter results in yield losses. In the present study, we have employed target specific amiRNA technology to engineer pod shatter resistance in crop Brassicas. In future, similar strategies can be outlined for redundantly silencing genes in various polyploid crop plants.

References

Ambesajir A, Kaushik A, Kaushik JJ, Petros ST (2012) RNA interference: A futuristic tool and its therapeutic applications. Saudi Journal of Biological Sciences 19:395-403.

Araújo P, Cesarino I, Carmello-Guerreiro SM, Dornelas MC (2013) *Citrus sinensis* L. Osbeck orthologs of FRUITFULL and SHATTERPROOF are differentially expressed during fruit development. Plant Growth Regulation 70:1-13.

Bandopadhyay L, Basu D, Sikdar SR (2013) Identification of genes involved in wild crucifer *Rorippa indica* resistance response on mustard aphid *Lipaphis erysimi* challenge. PLoS ONE 8:e73632.

Bennett EJ, Roberts JA, Wagstaff C (2011) The role of the pod in seed development: strategies for manipulating yield. New Phytologist 190:838-853.

Boss PK, Vivier M, Matsumoto S, Dry IB, Thomas MR (2001) A cDNA from grapevine (*Vitis vinifera* L.), which shows homology to AGAMOUS and SHATTERPROOF, is not only expressed in flowers but also throughout berry development. Plant Molecular Biology 45:541-553.

Bowman JL, Baum SF, Eshed Y, Putterill J, Alvarez J. 1999. 4 molecular genetics of gynoecium development in arabidopsis. In: Pedersen, R. A. and Schatten, G. P., editors. Current Topics in Developmental Biology. Academic Press. p. 155-205.

Chauhan JS, Singh KH, Kumar A (2006) Compendium of Rapeseed-mustard varieties notified in India. Directorate of Rapeseed-Mustard Research, Bharatpur, Rajasthan:

Child RD, Ulvskov P, Van Onckelen HA, Botterman J (1999) Strategies for engineering shatter resistance into oilseed rape pods. In Proceedings of 10 th International Rapeseed Congress, Cambridge.

Chung KS, Lee JH, Lee JS, Ahn aJH (2013) Fruit indehiscence caused by enhanced expression of NO TRANSMITTING TRACT in *Arabidopsis thaliana*. Mol. Cells 35:519-525.

Dardick C, Callahan AM (2014) Evolution of the fruit endocarp: molecular mechanisms underlying adaptations in seed protection and dispersal strategies. Frontiers in Plant Science 5:284.

Dastagiri MB, Gajula MNV, Patil GI (2014) World and Indian Agriculture: Revolutions & Multi Speed Strategies for future. Science Discovery 2: 14-26.

Dhakate P, Shivaraj SM, Singh A (2014) Design of artificial miRNA for redundant silencing of Brassica SHP1 and SHP2: transient assay-based validation of transcript cleavage from polyploid Brassicas. Acta Physiologiae Plantarum 36:2125-2135.

Dong Y, Wang Y-Z (2015) Seed shattering: from models to crops. Frontiers in Plant Science 6:

Duan C-G, Wang C-H, Fang R-X, Guo H-S (2008) Artificial MicroRNAs highly accessible to targets confer efficient virus resistance in plants. Journal of Virology 82:11084-11095.

Duan C-G, Wang C-H, Guo H-S (2012) Application of RNA silencing to plant disease resistance. Silence 3:5.

Favaro R, Pinyopich A, Battaglia R, Kooiker M, Borghi L, Ditta G et al. (2003) Mads-box protein complexes control carpel and ovule development in arabidopsis. The Plant Cell 15:2603-2611.

Ferrándiz C, Pelaz S, Yanofsky MF (1999) Control of carpel and fruit development in arabidopsis. Annual Review of Biochemistry 68:321-354.

Gill SS, Gill R, Kumar G, Pareek A, Sharma PC, Anjum NA et al. 2012. mustard: approaches for crop improvement and abiotic stress tolerance. Improving Crop Resistance to Abiotic Stress. Wiley-VCH Verlag GmbH & Co. KGaA. p. 1351-1368.

Giovannoni JJ (2004) Genetic regulation of fruit development and ripening. The Plant Cell 16:S170-S180.

Girin T, Stephenson P, Goldsack CMP, Kempin SA, Perez A, Pires N et al. (2010) Brassicaceae INDEHISCENT genes specify valve margin cell fate and repress replum formation. The Plant Journal 63:329-338.

Gu Q, Ferrandiz C, Yanofsky MF, Martienssen R (1998) The FRUITFULL MADS-box gene mediates cell differentiation during *Arabidopsis* fruit development. Development 125:1509-1517.

Gupta PC, Lal SK (2011) Heterosis and combining ability analysis for yield and its components in Indian mustard (*Brassica juncea* L. Czern & Coss). Acadmic Journal of Plant Science:45-52.

Helliwell CA, Waterhouse PM. 2005. Constructs and methods for hairpin RNA-mediated gene silencing in plants. Methods in Enzymology. Academic Press. p. 24-35.

Holzapfel C, Vera C, Phelps S, Nybo B (2010) Evaluating the effectiveness of pod-sealants for reducing shattering losses in several cultivars of straight combined canola. Final Project Report for the Saskatchewan Canola Development.

Jaradat MR, Ruegger M, Bowling A, Butler H, Cutler AJ (2014) A comprehensive transcriptome analysis of silique development and dehiscence in Arabidopsis and Brassica integrating genotypic, interspecies and developmental comparisons. GM Crops & Food 5:302-320.

Javed N, Tahir M, Geng J, Li G, McVetty PBE (2014) Identification of Brassica genotypes and molecular markers for increased seed oil content. Canadian Journal of Plant Science 94:1103-1108.

Jorgensen RA, Doetsch N, Müller A, Que Q, Gendler K, Napoli CA (2006) A paragenetic perspective on integration of RNA silencing into the epigenome and its role in the biology of higher plants. Cold Spring Harbor Symposia on Quantitative Biology 71:481-485.

Kim J, Somers DE (2010) Rapid assessment of gene function in the circadian clock using artificial microRNA in Arabidopsis mesophyll protoplasts. Plant Physiology 154:611-621.

Kuai J, Yang Y, Sun Y, Zhou G, Zuo Q, Wu J et al. (2015) Paclobutrazol increases canola seed yield by enhancing lodging and pod shatter resistance in *Brassica napus* L. Field Crops Research 180:10-20.

Kurihara Y, Watanabe Y (2004) Arabidopsis micro-RNA biogenesis through Dicer-like 1 protein functions. Proceedings of the National Academy of Sciences of the United States of America 101:12753-12758.

Levitin B, Richter D, Markovich I, Zik M (2008) Arabinogalactan proteins 6 and 11 are required for stamen and pollen function in Arabidopsis. The Plant Journal 56:351-363.

Liljegren S, Ditta G, Eshed Y, Savidge B, Bowman J, Yanofsky M (2000) SHATTERPROOF MADS-box genes control seed dispersal in *Arabidopsis*. Nature 404:766-770.

Liljegren SJ, Roeder AHK, Kempin SA, Gremski K, Østergaard L, Guimil S et al. (2004) Control of fruit patterning in *Arabidopsis* by INDEHISCENT. Cell 116:843-853.

Liu Q, Chen Y-Q (2010) A new mechanism in plant engineering: The potential roles of microRNAs in molecular breeding for crop improvement. Biotechnology Advances 28:301-307.

Lu R, Martin-Hernandez AM, Peart JR, Malcuit I, Baulcombe DC (2003) Virus-induced gene silencing in plants. Methods 30:296-303.

Lü S, Du X, Lu W, Chong K, Meng Z (2007) Two AGAMOUS-like MADS-box genes from *Taihangia rupestris* (Rosaceae) reveal independent trajectories in the evolution of class C and class D floral homeotic functions. Evolution & Development 9:92-104.

Macleod J (1981) Harvesting in oilseed rape. Cambridge Agricultural Publishing, Cambridge 107–120.

Maier B, Wendt S, Vanselow JT, Wallach T, Reischl S, Oehmke S et al. (2009) A large-scale functional RNAi screen reveals a role for CK2 in the mammalian circadian clock. Genes & Development 23:708-718.

Mandel MA, Yanofsky MF (1995) The *Arabidopsis* AGL8 MADS box gene is expressed in inflorescence meristems and is negatively regulated by APETALA1. The Plant Cell 7:1763-1771.

Meena PD, Thomas L, Singh D (2014) Assessment of yield losses in *Brassica juncea* due to downy mildew (*Hyaloperonospora brassicae*). Journal of Oilseed Brassica 5:73-77.

Michniewicz M, Zago MK, Abas L, Weijers D, Schweighofer A, Meskiene I et al. Antagonistic regulation of PIN phosphorylation by PP2A and PINOID directs auxin flux. Cell 130:1044-1056.

Molnar A, Bassett A, Thuenemann E, Schwach F, Karkare S, Ossowski S et al. (2009) Highly specific gene silencing by artificial microRNAs in the unicellular alga *Chlamydomonas reinhardtii*. The Plant Journal 58: 165-174.

Mühlhausen A, Lenser T, Mummenhoff K, Theißen G (2013) Evidence that an evolutionary transition from dehiscent to indehiscent fruits in Lepidium (Brassicaceae) was caused by a change in the control of valve margin identity genes. The Plant Journal 73:824-835.

Nanni L, Bitocchi E, Bellucci E, Rossi M, Rau D, Attene G et al. (2011) Nucleotide diversity of a genomic sequence similar to SHATTERPROOF (PvSHP1) in domesticated and wild common bean (*Phaseolus vulgaris* L.). Theoretical and Applied Genetics 123:1341-1357.

Nathan R, Muller-Landau HC (2000) Spatial patterns of seed dispersal, their determinants and consequences for recruitment. Trends in Ecology & Evolution 15:278-285.

Østergaard L (2009) Don't 'leaf' now. The making of a fruit. Current Opinion in Plant Biology 12:36-41.

Parizotto EA, Dunoyer P, Rahm N, Himber C, Voinnet O (2004) In vivo investigation of the transcription, processing, endonucleolytic activity, and functional relevance of the spatial distribution of a plant miRNA. Genes & Development 18:2237-2242.

Park JY, Koo DH, Hong CP, Lee SJ, Jeon JW, Lee SH et al. (2005) Physical mapping and microsynteny of Brassica rapa ssp. pekinensis genome corresponding to a 222 kbp gene-rich region of Arabidopsis chromosome 4 and partially duplicated on chromosome 5. Molecular Genetics and Genomics 274:579-588.

Park W, Zhai J, Lee J-Y (2009) Highly efficient gene silencing using perfect complementary artificial miRNA targeting AP1 or heteromeric artificial miRNA targeting AP1 and CAL genes. Plant cell reports 28:469-480.

Peng PF, Li YC, Mei DS, Colasanti J, Fu L, Liu J et al. (2015) Expression divergence of FRUITFULL homeologs enhanced pod shatter resistance in *Brassica napus.* Genetic and Molecular Research 14:871-885.

Pereira LG, Coimbra S, Oliveira H, Monteiro L, Sottomayor M (2006) Expression of arabinogalactan protein genes in pollen tubes of *Arabidopsis thaliana.* Planta 223:374-380.

Pesaresi P, Mizzotti C, Colombo M, Masiero S (2014) Genetic regulation and structural changes during tomato fruit development and ripening. Frontiers in Plant Science 5:124.

Pinyopich A, Ditta GS, Savidge B, Liljegren SJ, Baumann E, Wisman E et al. (2003) Assessing the redundancy of MADS-box genes during carpel and ovule development. Nature 424:85.

Priyamedha, Singh BK, Thomas L, Bala M, Singh VV, Singh D (2015) Status and perspective of canola quality rapeseed-mustard cultivation in India : a review. Journal of Oilseed Brassica 6:142-151.

Rajani S, Sundaresan V (2001) The *Arabidopsis* myc/bHLH gene ALCATRAZ enables cell separation in fruit dehiscence. Current Biology 11:1914-1922.

Raman H, Raman R, Kilian A, Detering F, Carling J, Coombes N et al. (2014) Genome-wide delineation of natural variation for pod shatter resistance in *Brassica napus*. PLOS ONE 9:e101673.

Ripoll JJ, Bailey LJ, Mai Q-A, Wu SL, Hon CT, Chapman EJ et al. (2015) microRNA regulation of fruit growth. Nature Plants 1:15036.

Sablok G, Pérez-Quintero ÁL, Hassan M, Tatarinova TV, López C (2011) Artificial microRNAs (amiRNAs) engineering – On how microRNA-based silencing methods have affected current plant silencing research. Biochemical and Biophysical Research Communications 406:315-319.

Schwab R, Ossowski S, Riester M, Warthmann N, Weigel D (2006) Highly specific gene silencing by artificial micrornas in Arabidopsis. The Plant Cell 18:1121-1133.

Schwab R, Ossowski S, Warthmann N, Weigel D. 2010. Directed gene silencing with artificial MicroRNAs. In: Meyers, B. C. and Green, P. J., editors. Plant MicroRNAs: Methods and Protocols. Totowa, NJ: Humana Press. p. 71-88.

Seymour G, Poole M, Manning K, King GJ (2008) Genetics and epigenetics of fruit development and ripening. Current Opinion in Plant Biology 11: 58-63.

Seymour GB, Østergaard L, Chapman NH, Knapp S, Martin C (2013) Fruit development and ripening. Annual Review of Plant Biology 64:219-241.

Shekhawat K, Rathore SS, Premi OP, Kandpal BK, Chauhan JS (2012) Advances in agronomic management of Indian Mustard (*Brassica juncea* (L.) Czernj. Cosson): An overview. International Journal of Agronomy 2012:1-14.

Spence J, Vercher Y, Gates P, Harris N (1996) 'Pod shatter' in *Arabidopsis thaliana, Brassica napus* and *B. juncea*. Journal of Microscopy 181: 195-203.

Squires TM, Gruwel MLH, Zhou R, Sokhansanj S, Abrams SR, Cutler AJ (2003) Dehydration and dehiscence in siliques of Brassica napus and *Brassica rapa*. Canadian Journal of Botany 81:248-254.

Tani E, Polidoros AN, Tsaftaris AS (2007) Characterization and expression analysis of FRUITFULL- and SHATTERPROOF-like genes from peach (*Prunus persica*) and their role in split-pit formation. Tree Physiol 27: 649-659.

Unver T, Budak H (2009) Virus-Induced Gene Silencing, a post transcriptional gene silencing method. International Journal of Plant Genomics 2009:

Van der Linden CG, Vosman B, Smulders MJ (2002) Cloning and characterization of four apple MADS box genes isolated from vegetative tissue. Journal of Experimental Botany 53:1025-1036.

Vrebalov J, Pan IL, Arroyo AJM, McQuinn R, Chung M, Poole M et al. (2009) Fleshy Fruit Expansion and Ripening Are Regulated by the Tomato SHATTERPROOF Gene TAGL1. The Plant Cell 21:3041-3062.

Xu P, Zhang Y, Kang L, Roossinck MJ, Mysore KS (2006) Computational estimation and experimental verification of off-target silencing during posttranscriptional gene silencing in plants. Plant Physiology 142:429-440.

Younis A, Siddique MI, Kim CK, Lim KB (2014) RNA interference (RNAi) induced gene silencing: A promising approach of hi-tech plant breeding. International Journal of Biological Sciences 10:1150–1158.

Zhao T, Wang W, Bai X, Qi Y (2009) Gene silencing by artificial microRNAs in Chlamydomonas. The Plant Journal 58:157-164.

Chapter - 4

Plant MicroRNA: Biogenesis and Role in Seed Development

—*Anirban Chakraborty, Subodh Verma and Sabhyata Bhatia*

4.1 Introduction

MicroRNAs (miRNAs) are endogenous noncoding RNAs that direct gene expression by targeting the protein coding messenger RNAs. These classes of small RNAs (approximately 20-24 nucleotides in length) were first reported in *Caenorhabditis elegans* (Ambros 2001). They regulate gene expression at post transcriptional level leading to phenomenon called post transcriptional gene silencing (PTGS), RNA interference (RNAi), co-supression and quelling (Romano and Macino 1992). miRNAs are processed from hairpin precursors called precursor miRNA (pre-miRNA), that releases both mature miRNA and its complementary miRNA* strands (Llave et al. 2002). The miRNA* strand is subsequently degraded leaving mature miRNA strand to bind to its target sequence. The mature miRNA strand can drive gene silencing either by degrading their target mRNAs or by translation inhibition. Plant miRNAs have sequences that closely match their target thus making it a case of perfect/near perfect complementation. Functionally plant miRNAs are known to be involved in regulation of various biological processes such as organ polarity, developmental timing, photoperiodism, vegetative & reproductive developments, hormone signaling and stress responses (Aukerman and Sakai 2003; Zhang et al. 2005; Khraiwesh et al. 2012; Nakabayashi et al. 2005).

The mode by which microRNAs control gene expression are now being manipulated in molecular biology research (Rhoades et al. 2002).

To understand interaction between miRNA and their targets, genome wide experimental approaches have been developed that includes miRNA overexpression, detailed promoter analysis of the miRNA coding genes and utilization of miRNA target mimics (Sharma et al. 2016). Various studies have evinced that miRNAs expressed in the early stage of embryogenesis inhibit genes from pre-maturation thereby preserving proper developmental stage. Null mutations in genes involved in miRNA biogenesis were also found to produced defective phenotypes(Nodine and Bartel 2010). These findings led to increase in curiosity among scientists to further elucidate the role of miRNA in seed development. Here we present some of the recent knowledge on the biogenesis of miRNA and their plausible role in seed development.

4.2 Biogenesis of miRNAs in plants

4.2.1 Transcription

miRNAs are transcribed from the genes coding for miRNA, that are mostly located in the intergenic regions (Reinhart et al. 2002)and appear as non coding Expressed sequence tag sites (ESTs) (Aukerman and Sakai 2003). RNA POL II is the major enzyme which transcribes the miRNA genes in eukaryotes. Interaction of transcriptional activator and mediator aids in binding of the RNA POLII to the promoter, leading to synthesis of primary miRNA transcripts (pri-miRNA). These pri-miRNAs are approximately 1kb in lengthand have the potential to form imperfect hairpin structures (Reinhart et al. 2002). The occurrence of high numbers of TATA and cis-regulatory motifs in MIR promoter sequences, indicates its regulation by many transacting factors (Megraw et al. 2006; Xie et al. 2005). Mediator 20A (MED20A), Mediator 18 (MED18), Mediator 17 (MED17), Not2a/2b (Negative on TATA less2) and Cell Division Cycle (CDC5) promotes MIR transcription. Whereas MED17, MED20A, and MED18 are subunits of conserved mediator complex, NOT2a/2b are core subunits of the conserved CARBON CATABOLITE REPRESSION4 (CCR4)-NOT complex. Mutants med 17, med20a, med18, cdc5 &not2a/2bcarrying loss of function mutation have been shown to be defective in miRNA processing pathway. CDC5 binds to the MIR promoters and MED

20A & Not2a/2b recruits RNA POL II to the promoters (Xie et al. 2005). Recent studies by (Li et al. 2017) have shown an indirect link between a ribosomal protein SHORT VALVE 1 (STV1) & MIR transcription. RNA POL II failed to occupy MIR promoters in *stv1-1* mutants, leading to failed transcription.

4.2.2 Processing

In the nucleus, the primary miRNA transcripts are subjected to 5′7-Methylguanosine capping and a 3′ polyadenylation tail which provides stability to the primary transcript (Zhang et al. 2005). The 5′cap composed of 7-methylguanosine is recognized by the cap binding protein complex (CBP 20 and CBP80) (Laubinger et al. 2008). *Arabidopsis* mutants *cbc20* and *cbc80/abh1* show severe defects in their pri-miRNA processing (Kim et al. 2008; Gregory et al. 2008). The pri-miRNAs are further processed to pre-miRNAs that are approximately 60nt in length and form irregular hairpin structures containing many mismatches in their base, loops in their internal regions, and bulges (Krol and Krzyzosiak 2006). In animals this processing is generally carried out by Drosha which uses Pasha (*C elegans & Drosophila*) / DGCR8 (in mammals) as a cofactor leading to formation of microprocessor complex (Yeom et al. 2006). Dicer carries out the step of cleaving and converting the pre-miRNA to miRNA-miRNA* duplex. However, in plants dsRNA specific DCL1 (Dicer like 1) is the only enzyme which helps in processing of both pri-miRNA to pre-miRNA and pre-miRNA to miRNA-miRNA* duplex. Meticulous identification of mature miRNA from pri-miRNA transcripts is driven by signals encoded by pri-miRNA structures that includes a 14-15 bp paired stem close to the origin of miRNA-miRNA* duplex, a paired stem of precise length far from the duplex and a terminal loop (Mateos et al. 2010; Werner et al. 2010).

DCL is an endoribonuclease possessing six domains each having specific roles in miRNA processing. Presence of these six domain types are characteristic features of animal, fungal and plant DCL proteins. These types include DExD-helicase, helicase-C, a conserved Duf283, a PAZ with a 3′ overhang binding pocket (Rogers and Chen 2013), RNase III and double stranded RNA-binding (dsRB) domains. The DExD domain is present in the N terminus whilst the helicase-C in the C terminus. Each DCL has two RNase III catalytic domains which binds to and cleaves dsRNA to miRNA (Margis et al. 2006). In plants the number of nucleotides present in a miRNA depends upon which member of the DCL family synthesizes it. DCL 1 and DCL4 generate 21 nts small RNA, whilst DCL2 and DCL4 generate 22 and 24 nts respectively (Rogers and Chen 2013). The interaction of DCL1 with the primary miRNA transcript is mediated by a nuclear protein called DAWDLE (DDL), that provides stabilization to the pri-miRNA (Yu et al. 2008). DDL is a forkhead-associated (FHA) domain containing protein that mediates protein–protein interactions in both prokaryotes and eukaryotes (Yu et al. 2008). *ddl* mutants carrying loss of function mutation in the DDL gene shows severe pleotropic defects in *Arabidopsis* (Yu et al. 2008). Recent reports have implicated involvement of another RNA binding protein PLEIOTROPIC REGULATORY LOCUS 1 (PRL1) in providing stabilization to the pri-miRNA immediately after their biogenesis (Li et al. 2017). In the nuclear dicing/ D bodies, DCL1 associate with HYPONASTIC LEAVES 1, SERRATE & C-TERMINAL DOMAIN PHOSPHATASE- LIKE 1 (CPL1) proteins and forms a microprocessor complex that processes both the pri/pre-miRNA (Manavella et al. 2012). HYL1 and SERRATE are required for canonical processing of the pri/pre miRNA (Manavella et al. 2012). A single stranded RNA binding protein TOUGH (TGH) partners with DCL1 complex and binds pri/pre-miRNA thus facilitating their processing. TGH possess a G-patch and a SWAP (Suppressor-of-White-A Pricot) domain and is thought to mediate interaction between pri-miRNA and HYL1 (Ren et al.,2012). *tgh-1* mutants containing T-DNA insertion in its 11th intron expressed reduced levels of mature miRNA in *Arabidopsis* (Ren et al., 2012). Employment of pri-miRNA to the processing complex is also mediated by STV1 which binds to the stem loop within pri-miRNA (Li et al.,2016).

HYPONASTIC LEAVES 1 (HYL1) possesses two tandem dsRNA binding domains, a nuclear localization signal at amino terminus and a protein-protein interaction domain at carboxy terminus. The N-terminal HYL1 fragment comprising two tandem dsRBDs recognize dsRNA and can facilitate pri-miRNA cleavage by DCL1

(Yang et al. 2010). Manavella et al. (2012) reported existence of DRB1/HYL1 protein mainly in phosphorylated state. A Mitogen activated protein kinase MPK3 phosphorylates specific serine residues (S42 & S159) present within the dsRNA binding domain of HYL1 in *Oryza sativa* and *Arabidopsis* thus regulating its activity (Raghuram et al. 2015). Phosphorylation by MPK3 negatively regulates miRNA accumulation consequently maintaining normal level of miRNA in (Raghuram et al. 2015). Optimal activity of HYL1 requires dephosphorylation of HYL1 which is accomplished by the phosphatase activity of C-TERMINAL DOMAIN PHOSPHATASE-LIKE 1 (CPL-1) which removes the phosphate group from the S42 & S159 of HYL1. CPL1 comprises of two dsRNA binding motif while its homolog CPL2 has one dsRNA binding (Manavella et al. 2012). The interaction between CPL1 & HYL1 is mediated by SERRATE (SE) which acts as a molecular scaffold (Rogers and Chen 2013). SE is a C_2H_2 Zinc finger which also helps in linking DCL1 & HYL1 to the Cap binding complex (Manavella et al. 2012) *se-1* mutants shows increased pri-miRNA levels and reduction in mature miRNAs, thus suggesting its role in processing of pri/pre-miRNA (Lobbes et al. 2006). Processing complex cleaves pre-miRNA to give rise to mature miRNA-miRNA* duplex where each strand in the duplex has two nucleotide overhangs at 3′ end (Chen 2005).

In diverse plant species 32 nucleotides of the miRNA-miRNA* duplex is 22-*O*-methylated by the S-adenosyl-methionine dependent methyltransferase HUA ENHANCER1 (HEN1) before it is transported out of the nucleus into the cytoplasm (Li et al. 2005). *HEN1* is necessary for generation of mature miRNA and the absence of methylation creates variation in size of miRNA due to combined effect of truncation and oligouridylation at the 3′end of the miRNA duplex (Li et al. 2005)

4.2.3 Transport

In animals, transport of pre-miRNA duplex from nucleus into the cytoplasm is driven by a Ran-GTP dependent dsRNA binding protein, Exportin-5. In *Arabidopsis* HASTY (HST), the homologue of Exportin 5 mediates the transfer of plant miRNAs (Park et al. 2005).

4.2.3.1 RISC assembly and role of Argonaute proteins

Argonaute protein serves as a major component of the RISC (RNA induced silencing complex) complex and contains PIWI domain and a PAZ domain. The PIWI domain resembles ribonuclease H whilst the PAZ domain binds to the 3′end of the mature miRNA. The two strands of miRNA-miRNA* duplex gets detached from each other and the mature miRNA strand gets loaded into an AGO protein. The strand that majorly contributes to the mature functional miRNA is called guide strand whereas the other one called the passenger strand is removed and subsequently degraded. The nucleotide present at the 5′end of the mature miRNA directs its sorting to different AGO proteins (Mi et al. 2008)where it forms a functional RISC.

Correct guide and passenger strand selection of miRNA is faciliated by HYL1 and CPL1 (Manavella et al. 2012). During the association of AGO1 with the miRNA/miRNA* duplex, HEAT SHOCK PROTEIN90 (HSP90) forms a complex with it (Iki et al. 2010). Removal of the passenger strand is mediated by disassociation of HSP90, and other proteins that are known to be associated with AGO1; SQUINT (SQN) being the most important of them (Iki et al. 2012). Binding of HSP90 to AGO1 triggers its chaperone activity which then induces a conformational change in AGO1. The sin qua non in cytoplasm of SQN(Prunet et al. 2008) and HSP90 suggests that AGO1 loading occurs in the cytoplasm (Iki et al. 2010).

4.2.3.2 Silencing

Sequencing of target mRNA degradome has thrown light on the slicing activity of plant miRNAs. The slicer endonuclease activity resides within PIWI domain of AGO proteins that directs cleavage of mRNA targets complementary to mature functional strand of the miRNA (Liu et al. 2004; Addo-Quaye et al. 2008). *Arabidopsis* AGO1, AGO2, AGO4, AGO7, and AGO10 have been reported to possess the slicer activity (Mi et al. 2008). The cleaved mRNA products are then either degraded or may contribute to *trans*-acting small interfering RNA (tasiRNA) biogenesis wherein the cleavage products of *TAS* transcript are transcribed by RNA-DEPENDENT RNA POLYMERASE6 to form dsRNA (Cuperus et al. 2010; Chen et al. 2010).

Translation repression is another mechanism of regulation wherein there is no significant change in the target mRNA level. In plants translation repression requires near perfect complementation between target and the (Gandikota et al. 2007; Brodersen et al. 2008). The exact mechanism by which plant miRNAs repress translation is not fully understood; however, association of miRNA with polysome in AGO1 dependent manner has been identified as the most probable mechanism. In addition to the known cleavage and translation repression, miRNA can direct gene silencing by DNA methylation too, thus regulating gene expression transcriptionally. For instance in rice, long miRNAs containing 24 nts processed by DCL-3 are sorted to AGO4 which then triggers methylation of cytosine residues at both *MIR* and target loci (Wu et al. 2010).

4.2.3.3 miRNA turnover

The universal law of nature also involves the degradation of the miRNAs once their function is done and they are no longer needed. This degradation is usually carried out by SMALL RNA DEGRADING NUCLEASES (SDN), a family of exoribonucleases that has been foundto degrade single-stranded small RNAs in *Arabidopsis* (Ramachandran and Chen 2008).

4.3 Role of miRNA in seed development

Role of miRNA in regulating seed development has been established primarily using reverse genetic approaches wherein mutants defective in the miRNA biogenesis pathway, with reduced or null miRNA activity, and (or) overexpressing the MIR encoding genes. Knocking down of plant miRNA involves various strategies; generation of artificial endogenous target mimics (eTMs), Short tandem target mimics (STTM), CRISPR-Cas being a few of them (Zhou et al. 2017; Zhang et al. 2017).

Deep analysis of the globular (8-celled) stage of embryo development in *dcl1* mutants showed upregulation of nine genes encoding transcription factors most of them being the targets of miRNA viz. miR156, miR159, miR160, miR166, miR319, miR824, miR168, miR393, miR400 and miR778 (Nodine and Bartel 2010). Amongst all, the most upregulated target was found to be *SPL10* (*Squamosa*

Promoter-Binding Protein-Like 10) and *SPL11*,well established targets of miR156. During the early stages of embryogenesis miR156 represses *SPL* which in turn prevents the premature expression of genes that are expressed during the maturation phase (Nodine and Bartel 2010). Overaccumulation of *SPL* transcripts during the globular stage not only resulted in defective cell divisions but also showed up-regulation of positive regulators of the seed maturation viz. *FUS3* (*FUSCA3*), *LEC1* (*LEAFY COTYLEDON1*) and *LEC2*. In addition the mutants also showed reduced levels of *ASIL1* (*ARABIDOPSIS 6B-INTERACTING PROTEIN1-LIKE1*). *ASIL* is a repressor of maturation regulators during embryogenesis and after germination; enhanced expression of *SPL* in *dcl1* mutants represses *ASIL1*/*ASIL2* and thereby precociously induces the maturation phase (Willmann et al. 2011). Furthermore, overexpression of miR166 represses the expression of two transcription factors named *PHABULOSA* (*PHB*) and PHAVOLUTA (*PHV*) which would otherwise transcriptionally activate *LEC2* and induce the seed maturation in early embryogenic stage (Tang et al. 2012). In contrast STTM156 plants expressing reduced levels of miR156 and increased levels of *SPL14* exhibited

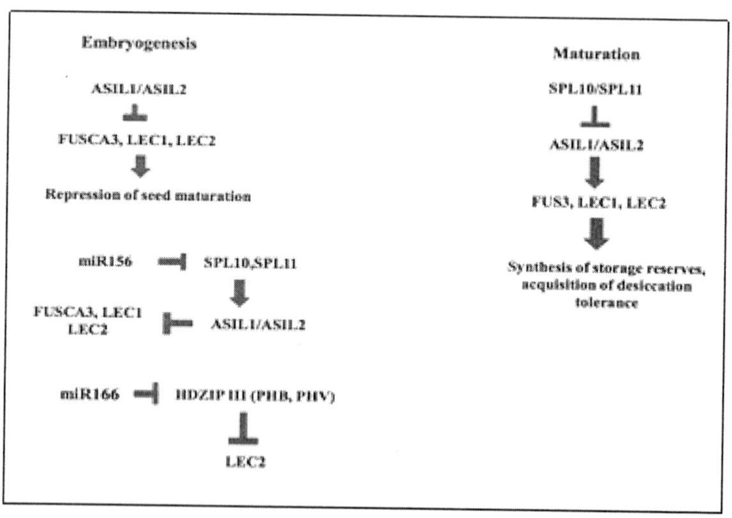

Figure 4.1: Models depicting the regulation of seed embryogenesis and maturation

enhanced grain number in rice panicle (Zhang et al. 2017). Taken together these results hint that the *miR156*/SPL network is quite complex and they go hand in hand with *miR166*/HD-ZIPIII during early embryo development to repress the embryonic maturation programme. A schematic representation of the aforesaid regulation is depicted in figure 4.1.

The basic features of embryogenesis entails establishment of axial and radial polarity. Proper patterning during early embryogenesis involves miR164 which tightly regulates five members of NAC (No Apical Meristem; ATAF1/2; and CUP-SHAPED COTYLEDON2). Overexpression of miR164 shows defective cotyledons which are either cup-shaped or are partially fused similar to those seen in *cuc1 cuc2* double mutants (Aida et al. 1997; Mallory et al. 2004)and *cuc1 and cuc2* single mutants(Aida et al. 1997; Vroemen et al. 2003) respectively. Plants with miR164-resistant versions of CUC1 shows deformity in cotyledon orientations, alterations in floral and vegetative development (Mallory et al. 2004) whereas those expressing miR164-resistant CUC2 proteins manifests enlarged boundary domains (Laufs et al. 2004), thus throwing light on the importance of miR164 for normal embryo development.

Patterning during embryogenesis also depends upon auxin concentration, their transport between tissues and their ability to regulate auxin-responsive (Bowman and Floyd 2008). Signal auxin triggers the transcription of Auxin Response Factors (ARFs), class of transcription factors that bind to auxin-responsive elements in the target promoters and regulate their expression either by activation or by repression (Abel and Theologis 1996). miR160 family targets such ARFs (ARF17), ARF16, and ARF10 (Rhoades et al. 2002) thereby maintaining auxin homeostasis that ensures proper patterning of the embryos. *Arabidopsis foc* (*floral organs in carpels*) mutant carrying Ds transposon in the 30 regulatory region of miR160a (Liu et al. 2010) show double-filled suspensors, asymmetric embryos at the globular stage, asymmetric cotyledons at the heart stage, and ultimately aberrant seeds due to abnormal cell division and differentiation. Such abnormal phenotypes can be attributed to the increased levels of ARF17, ARF16, and ARF10 (Rhoades et al. 2002), due to dramatic reduction in miR160a.

Another class of miRNA having indispensable role in seed development is miR172 which is highly conserved in higher plants and represses the expression of transcription factors containing APETALA2 (AP2) domains (Aukerman and Sakai 2003). In addition to specification of floral identity miR172 also helps in determining seed weight and yield in *Arabidopsis*. Loss-of function mutation in the AP2 gene greatly reduces male fertility thereby increasing the weight of seed (Jofuku et al. 2005).

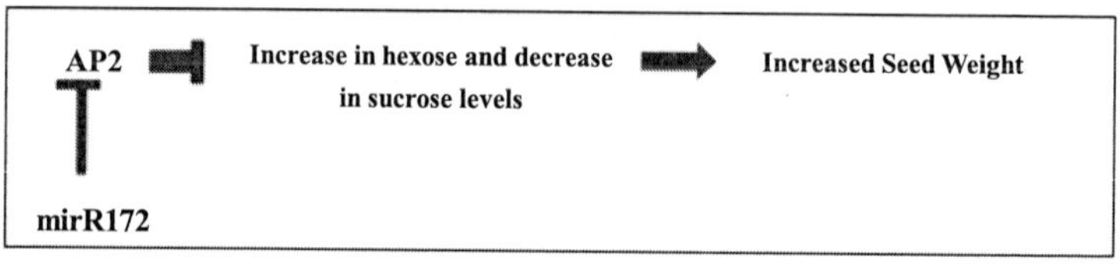

Figure 4.2: Model depicting the role of miR172 in regulating seed weight in plants

The increase in seed weight in loss of function *ap2* seeds can be attributed to an increase in cell size of the embryo (Jofuku et al. 2005; Ohto et al. 2005). Studies have shown that a high concentration of hexose strongly enhances the mitotic activity during the early embryogenesis whilst a higher sucrose to hexose aids in cell expansion and seed filling during the final maturation stage (Ohto et al. 2005). Mutation in the AP2 gene might have altered the hexose sucrose ratio resulting in enhanced cell divisions and thus an increased cell size. Taken together these results indicate that miR172 positively regulates seed size increase by suppressing AP2 (Figure 4.2).

4.4 Conclusion

RNA biology in plants saw a paradigm shift with miRNA directed regulation of trasncipt coming into spotlight. miRNA mediated regulation of transcripts have been found to be involved in regulating various aspects of development and stress responses. Development shift in plants is now being closely studied owing to its use in crop improvement; miRNA regulated processes such as seed development has likewise seen an upward trend (Zhang et al. 2017). With advancement of Next Generation Sequencing these studies have found their way into forward genetic for manipulation of crop traits.

References

Abel S, Theologis A (1996) Early genes and auxin action. Plant Physiology 111:9-17.

Addo-Quaye C, Eshoo TW, Bartel DP, Axtell MJ (2008) Endogenous siRNA and microRNA targets identified by sequencing of the Arabidopsis degradome. Current biology : CB 18:758-762.

Aida M, Ishida T, Fukaki H, Fujisawa H, Tasaka M (1997) Genes involved in organ separation in Arabidopsis: an analysis of the cup-shaped cotyledon mutant. The Plant Cell 9:841-857.

Ambros V (2001) microRNAs: Tiny regulators with great potential. Cell 107: 823-826.

Aukerman MJ, Sakai H (2003) Regulation of flowering time and floral organ identity by a microRNA and its APETALA2-like target genes. The Plant Cell 15:2730-2741.

Bowman JL, Floyd SK (2008) Patterning and polarity in seed plant shoots. Annual Review of Plant Biology 59:67-88.

Brodersen P, Sakvarelidze-Achard L, Bruun-Rasmussen M, Dunoyer P, Yamamoto YY, Sieburth L et al. (2008) Widespread translational inhibition by plant miRNAs and siRNAs. Science 320:1185.

Chen H-M, Chen L-T, Patel K, Li Y-H, Baulcombe DC, Wu S-H (2010) 22-nucleotide RNAs trigger secondary siRNA biogenesis in plants. Proceedings of the National Academy of Sciences of the United States of America 107:15269-15274.

Chen X (2005) microRNA biogenesis and function in plants. FEBS Letters 579:5923-5931.

Cuperus JT, Carbonell A, Fahlgren N, Garcia-Ruiz H, Burke RT, Takeda A et al. (2010) Unique functionality of 22-nt miRNAs in triggering RDR6-dependent siRNA biogenesis from target transcripts in Arabidopsis. Nature Structural & Amp; Molecular Biology 17:997.

Gandikota M, Birkenbihl RP, Höhmann S, Cardon GH, Saedler H, Huijser P (2007) The miRNA156/157 recognition element in the 32 UTR of the Arabidopsis SBP box gene SPL3 prevents early flowering by translational inhibition in seedlings. The Plant Journal 49:683-693.

Gregory BD, O'Malley RC, Lister R, Urich MA, Tonti-Filippini J, Chen H et al. (2008) A link between RNA metabolism and silencing affecting Arabidopsis development. Developmental Cell 14:854-866.

Iki T, Yoshikawa M, Meshi T, Ishikawa M (2012) Cyclophilin 40 facilitates HSP90-mediated RISC assembly in plants. The EMBO Journal 31: 267-278.

Iki T, Yoshikawa M, Nishikiori M, Jaudal MC, Matsumoto-Yokoyama E, Mitsuhara I et al. (2010) In vitro assembly of plant RNA-induced silencing complexes facilitated by molecular chaperone HSP90. Molecular Cell 39:282-291.

Jofuku KD, Omidyar PK, Gee Z, Okamuro JK (2005) Control of seed mass and seed yield by the floral homeotic gene APETALA2. Proceedings of the National Academy of Sciences of the United States of America 102:3117-3122.

Khraiwesh B, Zhu J-K, Zhu J (2012) Role of miRNAs and siRNAs in biotic and abiotic stress responses of plants. Biochimica et Biophysica Acta (BBA)-Gene Regulatory Mechanisms 1819:137-148.

Kim S, Yang J-Y, Xu J, Jang I-C, Prigge MJ, Chua N-H (2008) Two cap-binding proteins CBP20 and CBP80 are involved in processing primary microRNAs. Plant and Cell Physiology 49:1634-1644.

Krol J, Krzyzosiak WJ. 2006. Structure analysis of microRNA precursors. In: Ying, S.-Y., editor). MicroRNA Protocols. Totowa, NJ: Humana Press. p. 19-32.

Laubinger S, Sachsenberg T, Zeller G, Busch W, Lohmann JU, Rätsch G et al. (2008) Dual roles of the nuclear cap-binding complex and SERRATE in pre-mRNA splicing and microRNA processing in *Arabidopsis thaliana*. Proceedings of the National Academy of Sciences 105:8795-8800.

Laufs P, Peaucelle A, Morin H, Traas J (2004) MicroRNA regulation of the CUC genes is required for boundary size control in Arabidopsis meristems. Development 131:4311-4322.

Li J, Yang Z, Yu B, Liu J, Chen X (2005) Methylation protects miRNAs and siRNAs from a 32-end uridylation activity in Arabidopsis. Current Biology 15:1501-1507.

Li S, Liu K, Zhang S, Wang X, Rogers K, Ren G et al. (2017) STV1, a ribosomal protein, binds primary microRNA transcripts to promote their interaction with the processing complex in Arabidopsis. Proceedings of the National Academy of Sciences 114:1424-1429.

Liu J, Carmell MA, Rivas FV, Marsden CG, Thomson JM, Song J-J et al. (2004) Argonaute2 is the catalytic engine of mammalian RNAi. Science 305:1437.

Liu X, Huang J, Wang Y, Khanna K, Xie Z, Owen HA et al. (2010) The role of floral organs in carpels, an Arabidopsis loss-of-function mutation in MicroRNA160a, in organogenesis and the mechanism regulating its expression. The Plant Journal 62:416-428.

Llave C, Kasschau KD, Rector MA, Carrington JC (2002) Endogenous and silencing-associated small rnas in plants. The Plant Cell 14:1605.

Lobbes D, Rallapalli G, Schmidt DD, Martin C, Clarke J (2006) SERRATE: a new player on the plant microRNA scene. EMBO Reports 7:1052-1058.

Mallory AC, Dugas DV, Bartel DP, Bartel B (2004) microRNA regulation of nac-domain targets is required for proper formation and separation of adjacent embryonic, vegetative, and floral organs. Current Biology 14:1035-1046.

Manavella Pablo A, Hagmann J, Ott F, Laubinger S, Franz M, Macek B et al. (2012) fast-forward genetics identifies plant CPL phosphatases as regulators of miRNA processing factor HYL1. Cell 151:859-870.

Margis R, Fusaro AF, Smith NA, Curtin SJ, Watson JM, Finnegan EJ et al. (2006) The evolution and diversification of Dicers in plants. FEBS Letters 580:2442-2450.

Mateos JL, Bologna NG, Chorostecki U, Palatnik JF (2010) Identification of microRNA processing determinants by random mutagenesis of Arabidopsis mir172a precursor. Current Biology 20:49-54.

Megraw M, Baev V, Rusinov V, Jensen ST, Kalantidis K, Hatzigeorgiou AG (2006) MicroRNA promoter element discovery in Arabidopsis. RNA 12:1612-1619.

Mi S, Cai T, Hu Y, Chen Y, Hodges E, Ni F et al. (2008) Sorting of small RNAs into Arabidopsis argonaute complexes is directed by the 52 terminal nucleotide. Cell 133:116-127.

Nakabayashi K, Okamoto M, Koshiba T, Kamiya Y, Nambara E (2005) Genome-wide profiling of stored mRNA in *Arabidopsis thaliana* seed germination: epigenetic and genetic regulation of transcription in seed. The Plant Journal 41:697-709.

Nodine MD, Bartel DP (2010) MicroRNAs prevent precocious gene expression and enable pattern formation during plant embryogenesis. Genes & Development 24:2678-2692.

Ohto M-a, Fischer RL, Goldberg RB, Nakamura K, Harada JJ (2005) Control of seed mass by APETALA2. Proceedings of the National Academy of Sciences of the United States of America 102:3123-3128.

Park JY, Koo DH, Hong CP, Lee SJ, Jeon JW, Lee SH et al. (2005) Physical mapping and microsynteny of Brassica rapa ssp. pekinensis genome corresponding to a 222 kbp gene-rich region of Arabidopsis chromosome 4 and partially duplicated on chromosome 5. Molecular Genetics and Genomics 274:579-588.

Prunet N, Morel P, Thierry A-M, Eshed Y, Bowman JL, Negrutiu I et al. (2008) REBELOTE, SQUINT, and ULTRAPETALA1 function redundantly in the temporal regulation of floral meristem termination in Arabidopsis thaliana. The Plant Cell 20:901-919.

Raghuram B, Sheikh AH, Rustagi Y, Sinha AK (2015) MicroRNA biogenesis factor DRB1 is a phosphorylation target of mitogen activated protein kinase MPK3 in both rice and Arabidopsis. FEBS Journal 282:521-536.

Ramachandran V, Chen X (2008) Degradation of microRNAs by a family of exoribonucleases in Arabidopsis. Science (New York, N.Y.) 321:1490-1492.

Reinhart BJ, Weinstein EG, Rhoades MW, Bartel B, Bartel DP (2002) MicroRNAs in plants. Genes & Development 16:1616-1626.

Rhoades MW, Reinhart BJ, Lim LP, Burge CB, Bartel B, Bartel DP (2002) Prediction of plant microRNA targets. Cell 110:513-520.

Rogers K, Chen X (2013) Biogenesis, turnover, and mode of action of plant microRNAs. The Plant Cell 25:2383-2399.

Romano N, Macino G (1992) Quelling: transient inactivation of gene expression in *Neurospora crassa* by transformation with homologous sequences. Molecular Microbiology 6:3343-3353.

Sharma D, Tiwari M, Pandey A, Bhatia C, Sharma A, Trivedi PK (2016) MicroRNA858 Is a potential regulator of phenylpropanoid pathway and plant development. Plant Physiology 171:944-959.

Tang X, Bian S, Tang M, Lu Q, Li S, Liu X et al. (2012) MicroRNA–mediated repression of the seed maturation program during vegetative development in Arabidopsis. PLOS Genetics 8:e1003091.

Vroemen CW, Mordhorst AP, Albrecht C, Kwaaitaal MACJ, de Vries SC (2003) The CUP-SHAPED COTYLEDON3 gene is required for boundary and shoot meristem formation in Arabidopsis. The Plant Cell 15:1563-1577.

Werner S, Wollmann H, Schneeberger K, Weigel D (2010) Structure determinants for accurate processing of miR172a in *Arabidopsis thaliana*. Current Biology 20:42-48.

Willmann MR, Mehalick AJ, Packer RL, Jenik PD (2011) MicroRNAs regulate the timing of embryo maturation in Arabidopsis. Plant Physiology 155:1871-1884.

Wu L, Zhou H, Zhang Q, Zhang J, Ni F, Liu C et al. (2010) DNA methylation mediated by a microRNA pathway. Molecular Cell 38:465-475.

Xie Z, Allen E, Fahlgren N, Calamar A, Givan SA, Carrington JC (2005) Expression of Arabidopsis MIRNA genes. Plant Physiology 138:2145-2154.

Yang SW, Chen H-Y, Yang J, Machida S, Chua N-H, Yuan YA (2010) Structure of Arabidopsis HYPONASTIC LEAVES1 and its molecular implications for miRNA processing. Structure 18:594-605.

Yeom K-H, Lee Y, Han J, Suh MR, Kim VN (2006) Characterization of DGCR8/ Pasha, the essential cofactor for Drosha in primary miRNA processing. Nucleic Acids Research 34:4622-4629.

Yu B, Bi L, Zheng B, Ji L, Chevalier D, Agarwal M et al. (2008) The FHA domain proteins DAWDLE in Arabidopsis and SNIP1 in humans act in small RNA biogenesis. Proceedings of the National Academy of Sciences 105:10073-10078.

Zhang BH, Pan XP, Wang QL, Cobb GP, Anderson TA (2005) Identification and characterization of new plant microRNAs using EST analysis. Cell Research 15:336.

Zhang H, Zhang J, Yan J, Gou F, Mao Y, Tang G et al. (2017) Short tandem target mimic rice lines uncover functions of miRNAs in regulating important agronomic traits. Proceedings of the National Academy of Sciences 114:5277.

Zhou J, Deng K, Cheng Y, Zhong Z, Tian L, Tang X et al. (2017) CRISPR-Cas9 based genome editing reveals new insights into microRNA function and regulation in rice. Frontiers in Plant Science 8:1598.

Chapter - 5

Development and Applications of Molecular Markers in Jatropha

—*Pratima Pandey and Shashi Bhushan Tripathi*

5.1 Introduction

Energy security is an important policy issue all over the World. International energy markets have depended heavily on fossil fuels like crude oil, coal and natural gas, which provide almost 80 per cent of the world's supply of key energy. Being non-renewable in nature, the fossil fuels may become exhausted in the near future. Jatropha emerged as a biodiesel crop in past years. For genetic improvement of Jatropha several studies were carried out by researchers all over the world. Above chapter briefly discussed the level of genetic diversity present in Jatropha and analysis of genetic diversity using AFLP and microsatellite markers. Interspecific hybridization is also very common in Jatropha because of its high compatability with other species of Jatropha eg. *J. integerrimma.* Other important aspects like out crossing rates which is very important for hybridization were also discussed here along with lipid profiles in Jatropha species and their hybrids.

5.2 Genetic diversity studies in Jatropha

Analysis of genetic relationships in any plant species is an important component of species improvement program, as it serves to provide information about genetic diversity and for stratified sampling of breeding populations (Mohammadi and Prasanna 2003). Accurate assessment of the levels and patterns of genetic diversity can be invaluable in crop breeding for diverse applications including analysis of genetic variability in cultivars (Cox et al. 1986; Smith 1984) identifying diverse parental combinations to create segregating progenies with maximum variability for further selection (Barrett and Kidwell 1998) and introgressing desirable genes from diverse germplasm into the available genetic base (Thompson et al. 1998). In case of inbred lines or pure lines, an understanding of genetic relationships can be particularly useful in planning crosses by assigning lines to specific heterotic groups and for precise identification with respect to plant varietal protection (Hallauer and Miranda 1988).

For a conservationist and germplasm manager, analysis of genetic diversity in different accessions and germplasm collections can facilitate reliable classification of accessions and identification of core collections (Bamberg and del Rio 2014; Brown 1989). Identification of a core collection is an important step towards management and utilization of genetic resources. The core collection, a subsample of the whole collection, typically comprises about 10 per cent of all available accessions and is intended to provide a set of genetically diverse material (Brown 1989; Schoen and Brown 1993). A number of studies have used AFLP markers for identification of core collections due to their high abundance and ease of genotyping (Bamberg and del Rio 2014; Jansen and van Hintum 2007). An attempt has been made in the current work towards identification of core accessions in Jatropha using AFLP markers (Sinha et al. 2015c).

A large number of studies have been made so far to understand the genetic diversity in *J. curcas* using molecular markers. Basha and Sujatha (2007) used RAPD and ISSR markers to analyze the genetic diversity in set of 43 accessions of *J. curcas* from different geographical locations of India and 1 accession from Mexico. Out of 400 RAPD primers tested, 339 primers produced amplification products and 267 revealed polymorphic banding patterns. Six RAPD markers were specific to the Indian accessions while 17 markers were specific to the Mexican genotype. Out of 48 ISSR primers, 24 generated polymorphic banding patterns. Percent polymorphism of bands with ISSR primers ranged from 11.1 per cent to 66.6 per cent with an average polymorphism of 33.5 per cent. Two ISSR markers found in all the 42 Indian populations were absent in the Mexican genotype. Based on RAPD polymorphism the accessions diverged at 57 per cent variation while in case of ISSR the maximum genetic distance was 0.65 (Basha and Sujatha 2007).

Ganesh Ram et al. (Ganesh Ram et al. 2008) used RAPD markers to study the genetic diversity in 8 species of Jatropha including *J. curcas*. Out of 26 random primers used, 18 primers gave reproducible amplification banding patterns. Out of 134 bands scored across 12 accessions, 112 bands were polymorphic accounting for 80.2 per cent polymorphism. Jaccard's genetic similarity coefficient in the accessions varied from 0.00 to 0.85. The high genetic diversity in the dataset was expected due to the fact that the accessions belonged to different species. Such results also provide the possibilities of exploiting these species to broaden the genetic base of *J. curcas*. Many of the species used in the study such as *J. integerrima* and *J. podagrica* were found to be readily crossable with *J. curcas* especially when *J. curcas* was used as female parent (Dehgan 1984; Suwannoi et al. 2011).

Sun et al. (2008) employed 7 AFLP primer combinations and 17 microsatellite markers to study the genetic relationships of 58 *J. curcas* accessions from China. Interestingly, 14.3 per cent of the AFLP loci showed polymorphism while only one out of 17 SSR primers showed polymorphism with two alleles. In another study (Subramanyam et al. 2009) used RAPD markers on 40 accessions of *J. curcas* from India in which 94 (75.2%) out of 125 bands were polymorphic across the genotypes. Basha et al. (2009) used 17 microsatellite primers on a set of 72 *J. curcas* accessions from different countries (World collection). The study revealed variation in accessions from Mexico and El Salvador while accessions from other countries could not be distinguished.

In another study, Basha et al. (2009) carried out a comparative study of biochemical traits and several molecular markers (RAPD, ISSR and SSR) in 72 *J. curcas* germplasm from different countries including India. In this study, the RAPD primers revealed 42 per cent polymorphism in Indian accessions whereas 61.8 per cent polymorphism was observed in World collection indicating a higher genetic diversity in accessions from diverse geographical regions.

Interestingly, ISSR primers did not show greater polymorphism in this world collection (35.5 per cent) as compared to Indian accessions (34.1 per cent). This may be due to the differences in the target regions scanned by the two marker systems. RAPD markers scan both the coding as well as non-coding regions including repeats or single copy sequences. ISSR primers, on the contrary, detect polymorphisms in the microsatellite rich regions created due to DNA slippage.

Warrier and coworkers (2012) used allozyme and RAPD markers to assess genetic diversity of 56 accessions from India. Using data on 18 allozyme and 120 RAPD loci, they reported a low genetic diversity (H_e =0.205 with allozymes and 0.181 with RAPD). The data also indicated differentiation among populations. No significant association among genetic and geographical variables was observed suggesting that such low genetic variation in this allogamous species could be due to the few introductions that have spread across the country through vegetative propagation.

Biabani et al. (2013) used ISSR markers to assess diversity in 144 accessions belonging to 6 populations from Malaysia, Indonesia, the Philippines and India. Ten ISSR primers generated 143 (55 per cent) polymorphic fragments across the accessions. The coefficient of genetic differentiation (GST) was 0.13 which showed a medium level of differentiation of the populations. Majority of the genetic variation (87 per cent) was distributed within populations as compared to only minor (13 per cent) variation distributed between populations. The Malaysia1 and Indonesia1 populations were found to be the most distinct from each other.

Tatikonda et al. (2009) reported employed AFLP markers to assess the diversity in the elite germplasm collection of *J. curcas*. They used 48 accessions collected from six different states of India. Using 7 AFLP primer combinations, they scored a total of 770 fragments out of which 680 (88 per cent) showed polymorphism in the germplasm analyzed. They obtained 59 (8.7 per cent) unique (accession specific) fragments and 108 (15.9 per cent) rare (present in less than 10 per cent accessions) fragments. Majority of groups obtained in the phenogram and PCA contained accessions as per geographical locations. In general, accessions from Andhra Pradesh were found diverse as these were scattered in different groups, whereas accessions coming from Chhattisgarh had higher number of unique/rare fragments.

A comprehensive study involving a large germplasm collection of 192 *J. curcas* accessions across different regions of Brazil was carried out by Rosado et al. (2010). The study used 96 RAPD primers and six microsatellite markers. Only 23 of the 381 RAPD bands and one microsatellite marker were polymorphic. Surprisingly, all accessions were homozygous at all but one microsatellite, which was in contrast with the outcrossing mating system reported for the species. The results also suggested that *J. curcas* not only supports selfing but possibly breeds by geitonogamy. Similarity based clustering revealed only 43 unique multilocus profiles in the 192 accessions with a high probability of duplicate genotypes. No relationship between clustering of accessions and geographic origin was observed, suggesting that *J. curcas* has experienced a widespread dispersion across regions by seeds and possibly through vegetative propagules.

Zhang et al.(2011) analyzed genetic diversity of *J. curcas* accessions from China and other countries such as Mali, Burkina Faso, Indonesia, Burma and Laos. A total of 240 accessions were analyzed using 6 AFLP primer combinations/ Out of 352 scorable loci, 14.78 per cent were polymorphic. AMOVA results revealed that 36.18 per cent of the variance occurred among populations and 63.82 per cent occurred within populations. The study suggested that the genetic resources of *J. curcas* had a narrow genetic diversity in China and Southeast Asia. He et al. (2011) conducted AFLP analysis and concluded that genetic diversity is very limited in the Madagascar germplasm compared to the Mexican germplasm.

Na-ek et al. (2011) used 10 SSR markers on 32 accessions of *Jatropha curcas* and observed a low level of average genetic diversity ($H_e = 0.160$). Two major populations were detected by STRUCTURE. The analysis of molecular variance (AMOVA) showed higher variability (63.7 per cent) among groups than within groups (36.2 per cent) which is expected for a predominantly outcrossed species. Osorio et al. (2014) reported high level of molecular and phenotypic biodiversity in *Jatropha curcas* with SSR (Simple Sequence Repeat),

TRAP (Target Region Amplification Polymorphism) and AFLP (Amplified fragment length polymorphism) markers. The pool of Central American accessions showed very large genetic variation and the highest phenotypic variation compared to Asian, African and South American accessions and should therefore be considered as the most important source for plant breeding.

5.3 Inter-specific hybridization and development of linkage map in Jatropha

Narrow genetic diversity within *J. curcas* gene pool has been identified as a common problem in most of the published studies mentioned in earlier section. Interspecific hybridization is a logical and potential strategy to widen the gene pool of any species (Stalker 1980). Linkage maps are valuable for a wide range of applications, including the identification and positional cloning of important genes (Martin et al. 1993) and the implementation of marker-assisted selection (Collard and Mackill 2008). Mapping populations such as F2 and BC1 generated through interspecific crosses have been used for construction of linkage maps and identification of important loci in different crops. Blum et al. (2002) used F2 populations of *Capsicum annum x C. frutescens* to map the pungency locus in chili. Yi et al. (2006a) used F2 populations of *Capsicum annum x C. frutescens* for linkage mapping using microsatellite markers. A number of efforts have been made in the past towards interspecific hybridization in Jatropha.

Interspecific hybridization is a useful method to increase genetic diversity and generating prebreeding material for crop improvement. Further, mapping populations generated through interspecific hybridization are more efficient in development of linkage maps due to high polymorphism between the parental genotypes. In recent years, research on this aspect has increased in Jatropha.

Rupert et al (1970) obtained viable hybrid seeds from a series of crossing experiments between *J. curcas* and *J. integerimma* using *J. curcas* as the female parent. These seeds produced plants which were intermediate between the parental species in a number of characteristics. No seed could, however, be obtained from the reciprocal cross (*J. integerimma* x *J. curcas*). Dehgan (1984) studied interspecific hybridization compatibilities among 11 Jatropha species. He investigated different types of compatibility barriers between different species. He also described the phylogenetic significance of interspecific hybridization in Jatropha expressed by number of flowers pollinated, number of seeds obtained, and percentage germination, color configuration of sepals, petals and leaves. He also studied different types of compatibility barriers between different species.

A natural hybrid between *J. curcas* × *J. gossypifolia,* named as *J. tanjorensis*, was reported from some districts of Tamil Nadu (Prabakaran and Sujatha 1999).No fruit set was noticed in this species which was presumably due to abnormal meiotic division in the hybrid resulting from the parental genomic differences (Prabakaran and Sujatha 1999). Dehgan (1984) had also attributed the failure of interspecific hybrid between these two species due to their high phylogenetic distance. In India, interspecific hybridization between *J. curcas* and *J. integerrima* and subsequent backcrossing resulted in a number of flower colours varying from dark pink through green to white which had potential ornamental value (Sujatha and Prabakaran 2003).

Parthiban et al. (2009) also obtained successful hybrids between *J. curcas* and *J. integerrima*. All other interspecific crosses failed to produce F1 seeds due to existence of pre- and postzygotic incompatibility barriers in pollen-pistil interaction leading to problems in pollen tube growth and fertilization (Kumar et al. 2009). The F1 hybrids between *J. curcas* and *J. integerrima* exhibited vigorous growth, but the fruits were small in size indicating *J. integerrima* characters. Backcross progenies showed unique characteristics of fruit, seed and oil yield. Dhillon and coworkers (2009) developed F1 hybrids between *J. curcas* x *J. integerrima* and validated them by using RAPD markers. The hybrids resembled the female parent in terms of stem and leaf type while they resembled the male parent for leaf pigmentation. In another study, several BC_1 genotypes with low phorbol esters were identified (Popluechai et al. 2009).

Laosatit et al. (2014) also reported interspecific hybridization in Jatropha species and confirmed their hybridity using expressed sequence tag (EST)–simple sequence repeat markers developed from the *J. curcas* EST database. It is interesting to note that almost all successful interspecific crosses reported so far in *J. curcas* involved *J. integerrima* as the male parent. The high success of this combination is likely due to the demonstrated greater genetic similarities between the two species based on morphological (Dehgan 1984; Dehgan and Webster 1979), nuclear ITS (Pamidimarri et al. 2009a), RAPD (Ganesh Ram et al. 2008; Pamidimarri et al. 2009b) and AFLP data (Sinha et al. 2014; Sinha and Tripathi 2013) as compared to other species.

Two studies have reported linkage mapping in Jatropha so far (King et al. 2013; Wang ct al. 2011). The first linkage map (Wang et al. 2011) used a BC_1 population derived from interspecific cross between *J. curcas* and *J. integerrima.* The map contained 216 EST-SSR and 290 SNP markers spanning a total of 1440 cM with average marker to marker spacing of 2.8 cM ranging from 1.2 to 4.3 cM. The span of 11 linkage groups (LG) ranged from 84.9 to 187.5 cM. The number of markers on each linkage group ranged from 22 for LG 5 to 36 for LG6. LG 11 possessed the highest density of markers with marker spacing shorter than 2 cM, whereas LGs 1, 4 and 8 had the relatively lower density of markers with marker spacing greater than 4 cM. A comparative map between *J. curcas* and *A. thaliana* was also developed which contained192 loci derived from expressed sequence tags (ESTs). In another study, these genetic resources and tools were used to investigate the expression of oleosin genes in mature seeds (Liu et al. 2011). SNP markers were developed in from 3 oleosin genes, namely, *Ole*I, *Ole*II and *Ole*III, which were mapped on linkage groups 5, 3 and 8, respectively. Composite interval mapping identified 18 QTLs dispersed among all the linkage groups except LGs 3 and 11 controlling the oil traits in Jatropha. Four QTLs controlling total oil content were detected. The most effective QTL was spotted on LG4 explaining 11.1 per cent of the variation.

Sun and coworkers (Sun et al. 2012)identified a total of 28 QTLs for 11 growth and seed related traits using a BC_1 population of 296 individuals. A linkage map was constructed using 105 SSR (simple sequence repeat) markers dispersed on 11 linkage groups. Two QTLs, qTSW-5 and qTSW-7 controlling seed yield were mapped on LGs 5 and 7 respectively, where two QTL clusters controlling yield related traits were detected harboring five and four QTLs respectively. The favored alleles for different traits came from both the parents which indicate that pyramiding of positive alleles from the two parents would be required to achieve the ideal genotypes for high productivity.

King et al. (2013) used intraspecific F2 populations for linkage mapping and identification of QTLs regulating phorbol ester (PE) content in seeds. They used 4 F_2 populations and used SSR markers to develop an integrated map. The integrated map contained 502 SSR loci spanning a total distance of 717 cM. In one of the mapping populations (G33 × G43), a locus controlling phorbol ester biosynthesis was identified at 41 cM on linkage group 8 of the linkage map. Further analysis of the data confirmed that all F2 individuals which were homozygous for the G43 (nontoxic) allele at this locus yielded seeds which did not contain PE. Interestingly, the mean values for the PE content of the F2 plants which were heterozygous at this locus was also significantly less than the plants which were homozygous for the G33 (toxic) allele.

Two genome sequencing projects in Jatropha have been reported so far. One of these has been made publically available at www.kazusa.or.jp/jatropha (Hirakawa et al. 2012; Sato et al. 2010). The size of this assembled genome data is 297 Mbp. Another draft genome sequence of Jatropha has recently been released with a total length of 320.5 Mbp and containing 27,172 putative protein-coding genes (Wu et al. 2015). They also established a linkage map containing 1,208 markers, and anchored the genome assembly (81.7%) to this map to produce eleven pseudo chromosomes. The size of both the publically available genome sequences are consistent with the earlier reported estimated size of 410 Mbp using flow cytometry (Carvalho et al. 2008). In addition, another genome sequencing project was carried out jointly by Life Technologies and SG Biofuels but the same has not been made publically available (Accessed on January 30, 2015, at https://www.genomeweb.com/sequencing/life-technologies-and-sg-biofuels-sequence-jatropha-genome-solid-4).

5.4 Development of microsatellite markers in Jatropha

Microsatellites are tandem repeats of 1-6 nucleotides found at high frequency in the genomes of higher taxa. A good resource of microsatellite markers in Jatropha has been generated and is now publicly available including this study (Sinha et al. 2015b). However, no microsatellite based marker resource was available when the current work was initiated. Sun et al. (2008) developed 17 microsatellite markers using FIASCO (Fast Isolation by AFLP of Sequences Containing repeats) protocol and used them to study the genetic relationships of 58 *J. curcas* accessions. Only one SSR primer was found to be polymorphic with two alleles.

Pamidimarri et al. (2009c) isolated 12 microsatellites from *J. curcas* and tested their cross-species amplification in six species of Jatropha. Within *J. curcas*, observed and expected heterozygosities ranged from 0.94 to 0.54 and from 0.95 to 0.56, respectively. There was no significant linkage disequilibrium detected between any of the loci. The same group reported isolation of novel microsatellites in Jatropha by using FIASCO (Fast isolation by AFLP of sequences containing repeats) method (Pamidimarri et al. 2010).The overall efficiency of the enrichment of microsatellites by dual probe in was 54 per cent and the percentage of sequences having suitable flanking regions for the primer designing was 89.58 per cent among the isolated sequences. They observed that the overall diversity revealed by microsatellite markers was low in comparison to the diversity reported by multilocus marker systems. In another study, the investigators checked the ability of cross species amplification of microsatellite markers isolated from Jatropha curcas (Pamidimarri et al. 2011). Out of the 49 markers, 31 markers showed cross species amplification in all the species studied namely, *J. glandulifera, J. gossypifolia, J. integerrima, J. multifida, J. podagrica,* and *J. tanjorensis*.

Wen and others (2010) studied the transferability of 419 expressed sequence tag (EST) SSRs and 182 genomic SSRs from another Euphorbiaceae plant, cassava (*Manihot esculenta*) to Jatropha. In total, 187 out of 419 expressed sequence tag (EST)-SSR and 54 out of 182 genomic (G)-SSR markers from cassava were polymorphic among the *J. curcas* accessions. Thirty-six EST-SSRs and 20 G-SSRs detected a total of 183 polymorphic alleles in 45 *J. curcas* accessions. The clustering of accessions showed correlation with their geographic origin. The estimated mean genetic diversity index was 0.5572 suggesting a high level of genetic diversity in the germplasm collection. Sharma and Chauhan (2011) identified 5,80,986 SSRs from sequence analysis of another Euphorbiaceae plant, castor (*Ricinus communis*). Genomic distribution of SSRs revealed that 27 per cent were present in the non-genic region whereas 73 per cent were also present in the putative genic regions. Out of 302 randomly selected SSRs, 211 (~70 per cent) showed amplification products and 7.58 per cent showed polymorphisms within *J. curcas* genotypes. Tanya and coworkers (2011) developed 49 novel microsatellite markers based on EST sequences of Jatropha from NCBI databases and identified 8 polymorphic markers which could differentiate the toxic and non-toxic varieties of Jatropha. In another study, 11 out of 55 markers showed polymorphism, when evaluated on 26 accessions of *J. curcas* collected from several provinces in Thailand (Phumichai et al. 2011). Each marker showed 2 to 5 alleles with average polymorphic information content (PIC) of 0.49.

Yadav et al. (2011) designed 406 SSR primer pairs from EST-SSR sequences. Fifty randomly selected EST-SSR markers were amplified in 25 accessions collected from different geographical regions of India which revealed 21 polymorphic markers with 2 to 4 alleles with a total of 51 alleles. Polymorphic information content values ranged between 0.04 and 0.61 with an average of 0.25±0.16, indicating low to moderate level of informativeness within these EST-SSRs. The polymorphic markers showed 57.0 per cent to 95.6 per cent transferability among five species of *Jatropha* and 47.0 per cent transferability in *Ricinus communis*. Bressan and coworkers (2012) developed 40 microsatellites in Jatropha through enrichment of genomic libraries and evaluated their transferability in other species of family Euphorbiaceae.

5.5 Estimation of out crossing rates in *Jatropha curcas*

The out crossing rate is defined as the proportion of out crossed progeny produced by an individual mother or population. Muller (1932) stated that the primary function of sexual reproduction is to maintain and increase genetic diversity in populations. Reproductive phase is, therefore, very important in flowering plants as it represents the only opportunity for genes to move among conspecific populations (Levin and Kerster 1974). The mating system, as a part of the complete genetic system of the species, defines the mode of the transmission of genetic information from the parental to the next filial generation and is an important determining factor of plant population structure (Stern and Roche 1974). A species can produce offspring through different types of mating, such as random outcrossing, correlated outcrossing, biparental inbreeding, self-pollination, apomixes or a combination of these (Santos and Gama 2013). In general, outbreeding plants show higher levels of genetic diversity and less differentiation among each other than populations consisting of inbreeding plants (Allard et al. 1968; Brown 1979). The majority of the genetic variability in case of outcrossed species is distributed within the population. On the other hand, in predominantly self-pollinated species, majority of the genetic variability is distributed between populations (Hamrick and Godt 1989). A majority of tree species are allogamous or possess a mixed mating system that is predominantly allogamous (Gusson et al. 2006).

There are several factors which influence the mating system. Dioecy, androdioecy, gynodioecy (Wolff et al. 1988), protandry (Kesseli and Jain 1985; Schoen 1982) or protogyny (Bos et al. 1985) and the presence of self-incompatibility systems (Ennos 1985) are some of the mechanisms which promote outcrossing. Plant density (Schmitt 1983; Watkins and Levin 1990), foraging behavior of the pollinators (Levin and Kerster 1969), flower colour (Epperson and Clegg 1987) and morphology (Marshall and Abbott 1982) are some of the additional environmental factors influencing the pollen flow. Smaller populations promote high levels of inbreeding and random genetic drift (Frankel and Soule 1981; Franklin 1980). Fitness of individuals is affected by inbreeding depression and loss of alleles which in turn increases the probability of extinction for the population (Frankel and Soule 1981; Schonewald- Cox et al. 1983).

The effective population size is highly influenced by mating system and decreases with increasing levels of self-fertilization (Barrett and Kohn 1991). It means that the rate of outcrossing influences the rate of loss of genetic variation present in the plant species. Wright (1965) gave the estimation of fixation indices for the adult populations in which $F_{IS} = 1 - H_o/H_e$ where, H_o and H_e are the observed and expected frequency of heterozygotes. Outcrossing rates have been traditionally studied using codominant markers such as isozymes and microsatellites due to their high allelic polymorphism (Adugna et al. 2013; Chaix et al. 2003). However, dominant markers such as RAPD and AFLP have also been used successfully especially when microsatellite markers were unavailable (Gaiotto et al. 1997; Muluvi et al. 2004). Although these dominant markers provide less information per locus as compared to codominant markers, this limitation can be readily overcome by multilocus estimation of outcrossing with dominant markers having intermediate gene frequencies (Ritland and Jain 1981).

Biparental inbreeding, or mating between relatives, causes apparent selfing or increased homozygosity, relative to random mating. When true selfing is also present, the difference between multilocus and single-locus estimates of outcrossing, (t_m - t_s), is often used to characterize the level of biparental inbreeding (Ritland and Jain 1981). This difference is positive because single-locus estimates includes all apparent selfing due to biparental inbreeding, whereas multilocus estimates exclude much of the apparent selfing due to biparental inbreeding, as an observed outcross at any locus overrides the apparent selfing manifested at other loci. Thus, if inbreeding other than selfing occurs, t_s will generally underestimate outcrossing to a much greater extent than t_m. Therefore, estimates obtained using larger numbers of loci are closer to actual values.

This most commonly used model for analysis of mating systems is mixed mating model (Ritland and Jain 1981), which has been implemented in the program MLTR (Ritland 2002). The model assumes that a fraction of progeny are derived from self-fertilization and the remainder derived from outcrossing at random. From progeny array data, the program simultaneously estimates (i) the multilocus outcrossing rates (t_m) by the Newton-Raphson method; (ii) the mean single-locus outcrossing rate (t_s); (iii) the correlation of out-crossed paternity (r_p); (iv) correlation of selfing among loci (r_s); (v) single-locus inbreeding coefficient (Wright's fixation index) of the maternal parents (F); (vi) the pollen and ovule allele frequencies by the expectation-maximization method and; (vii)variances of the above quantities using the bootstrap method. The correlation of paternity (r_p) is the probability that the two siblings are outcrossed full-sibs. The correlation of selfing among loci, r_s, is the probability that any two loci are selfed. The value of r_s lies between -1 and 1, with values less than zero not possible under any realistic scenarios. A value of 1 indicates only uniparental inbreeding and no biparental inbreeding.

Gaiotto et al. (1997) used a combination of two dominant markers, i.e., AFLP and RAPD for estimation of outcrossing rates in *Eucalyptus urophylla* open pollinated progenies. A slightly higher multilocus outcrossing rate was obtained with RAPD (0.93±0.027) than AFLP markers (0.89± 0.033). Muluvi et al. (2004) studied outcrossing rates in *Moringa oleifera* using dominant AFLP markers. The multilocus outcrossing rate (t_m) was 0.74 with a significant level (22 per cent) of mating among relatives (biparental inbreeding). The results clearly indicated that *M. oleifera* is self-compatible and breeding schemes like inbred lines can be used provided the inbreeding depression is weak.

In *Sorghum bicolor,* the multilocus outcrossing rates ranged between31 to 65 per cent among populations, whereas within-family outcrossing rates were in the extreme range of 0 to 100 per cent. The highest outcrossing (tm = 0.65) was recorded in a weedy population that was intermixed with an improved sorghum variety. Thus, the outcrossing rates may be influenced by variations among the individuals at different marker loci as they help in detecting the outcrossing events.

J. curcas is monoecious but bears distinct male and female flowers which are arranged in a dichasial cyme. Flowering begins as early as 4 months after germination and but may take more than a year in some cases (Fresnedo-Ramirez 2013; Heller 1996). The female flowers are comparatively larger in size and are situated at the terminal positions of different branches of the cyme. In some cases, male flowers develop at the expected female flower positions. Flowering within a cyme is generally asynchronous with flowers on main axis opening earlier than those on sub-axes. As the flowers are unisexual, fruit set is only through geitonogamy or xenogamy (Raju and Ezradanam 2002). Fruit set through apomixes has also been reported (Bressan et al. 2013). Male flowers are in much higher numbers in an inflorescence as compared to female flowers. Pranesh et al. (2010) reported that the average number of female and male flowers per inflorescence was 14.4 (9.6-17.82) and 105.73 (75-138.07) respectively. Within the cyme, both protandry (shedding of pollen before stigma receptivity) and protogyny has been reported in Jatropha (Heller 1996; Raju and Ezradanam 2002). Nectar production of both male and female flowers peaked in the morning, declined in the afternoon, and remained low during night in all of their anthesis days (Luo et al. 2011). The common pollinators are Hymenopteran and Coleopteran insects such as *Apis dorsata* and *Apis mellifera*and nocturnal visitors such as moths which mainly fed on pollen and nectar. Luo et al. (2012) found that ants were the most abundant pollinators, accounting for 71.03 and 78.17 per cent of total visits at two study sites. Although the flowers exposed only to the ants produced less fruit than those exposed only to the winged visitors, ants alone resulted in almost 60 per cent fruit set.

J. curcas has been described as an outcrossed species. The pollen biology of *J. curcas* as described above also supports cross pollination (Ahoton and Quenum 2012; Kaur et al. 2011; Raju and Ezradanam 2002). However, very few efforts have been made to understand the mating system in *J. curcas* using progeny arrays (Sinha et al. 2013).Recently, a study on outcrossing rates in Jatropha using molecular markers has been published (Bressan et al. 2013).The multilocus outcrossing rates (t_m)ranged from 0.21 to 1.0 with a mean t_m of 0.683, showing that the species exhibits a mixed mating system. A significant level of biparental inbreeding (0.347) as expressed by difference between single and multilocus outcrossing rates was observed which suggested that mating among related individuals is quite prevalent in this species. Interestingly, 13 per cent of the offspring displayed an identical multilocus genotype to the mother tree which was attributed to apomixis. Apomictic fruit set and seed production has been reported in *J. curcas* in artificial pollination experiments under field conditions by protecting female flowers. Apomixis rates of 32 to 36 per cent have been reported from field grown *J. curcas* in India (Bhattacharya et al. 2005) and China (Chang-Wei et al. 2007).

However, a much lower level of apomixis (5 per cent) was reported from a study in Brazil (Bressan et al. 2013).

5.6 Oil content and lipid profiles in Jatropha

Biodiesel consists of the methyl esters of fatty acids obtained through transesterification of oil. Plant oils are triacylglycerols (TAGs) consisting of three fatty acid chains (16-18 carbon long) esterified to glycerol. The biodiesel properties of oil such as viscosity, cetane number, flash point, cloud point and stability depend on its fatty acid composition. Glycerol is mainly synthesized from glucose which is converted to fructose-1, 6- bisphosphate and then to dihydroxyacetone phosphate (DHAP) via glycolysis pathway. DHAP is converted to glycerol-3-phosphate by glycerol-3-phosphate dehydrogenase (G3pdh) and finally to glycerol by glycerol-3-phosphate phosphatase (G3pp) (Wang et al. 2001). All these reactions take place in the cytoplasm.

In higher plants, the fatty acids are synthesized *de novo* within the plastids. The synthesis begins with carboxylation of acetyl-CoA to malonyl-CoA by acetyl-CoA carboxylase (ACCase) in a 2-step reaction. Consequently, from malonyl CoA, a pool of mostly 16:0-

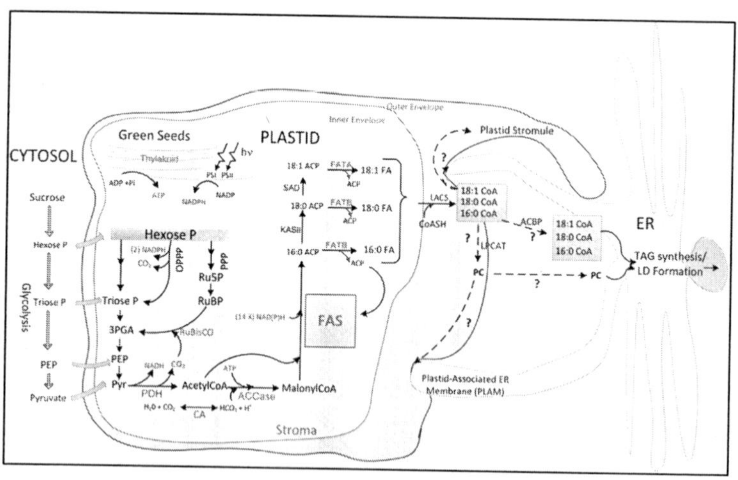

Figure 5.1: Fatty acid biosynthesis pathway. Adapted from Pollard and Ohlrogge (1999)

and 18:0-acyl chains (linked to acyl carrier protein or ACP) are produced through a series of condensation and elongation reactions. A portion of these acyl chains are also desaturated to form monounsaturated acyl-ACPs. These acyl-ACPs are cleaved by specific thioesterases and converted to acyl-CoAs by acyl-CoA synthetases at or near the outer plastid envelope. The acyl-CoAs are then exported to endoplasmic reticulum where they may be further elongated or incorporated into triacylglycerols (Pollard and Ohlrogge 1999).

The oil content in *J. curcas* seeds varies between 24 to 40 per cent (Kaushik et al. 2007). The oil content in other species of Jatropha such as *J. glandulifera, J. gossypifolia* and *J. multifida* has been found to be much lower than that in *J. curcas* (Banerji et al. 1985). Ahmed and Salimon (2009) reported that Malaysian Jatropha samples had highest crude oil percentage of 33.73 per cent than the Indian and Indonesian samples. Phorbol ester content was also low in Malaysian seeds (0.23 per cent) whereas it was high in Indonesian (1.58 per cent) and Indian (0.58 per cent) respectively.

A significant proportion of the reported variation in oil content is due to variations in seed quality which could be environmental. Variation in oil content could also be due to the process of oil estimation used, namely solvent extraction and cold press method. Solvent extraction with hexane is able to extract most of the oil content whereas, cold press can extract only about 75 per cent of the total oil present in the seeds (Gunstone 2007).

The Jatropha oil contains about 75 per cent unsaturated fatty acids, namely oleic acid and linoleic acid (Biello 2009; Desai and Vyas 1949). The major fatty acids in Jatropha oil are oleic acid, linoleic acid, palmitic acid and stearic acid (Naik et al. 2008). Banerji et al. (1985) found that oil samples of *J. curcas, J. glandulifera, J. gossypifolia* and *J. multifida* are rich in unsaturated fatty acids, namely, oleic and linoleic acid (72.2-84 per cent). The energy value of *J. multifida* oil was the highest (13.647kcal/g). Becker and Makkar (2008) compared the fatty acid composition of toxic and nontoxic varieties of *Jatropha curcas*. They found that palmitic (15.3 per cent) and oleic acid (41 per cent) were high in the toxic variety while linoleic was (42.1 per cent) high in the non-toxic variety. Liu et al. (2011) compared oil composition of *J. curcas* and *J. integerrima* and found that *J. curcas* has a higher proportion of oleic acid and lower proportion of linoleic acid as compared to *J. integerrima*.

Jatropha oil can be converted into fatty acid methyl esters (FAME) by transesterification using methanol in the presence of an alkali catalyst. The fatty acid methyl esters are then blended in different proportions with petro-diesel to get the final product (Karmee and Chadha 2005). The biodiesel properties of Jatropha oil have been shown to be within the limits of American Society for Testing and Materials (ASTM) specifications for diesel, D 6751 (Kywe and Oo 2009).

Several parameters are used to describe the biodiesel properties of oil (Biodiesel Standard 2002). Saponification number (SN) indicates the average chain length of all the fatty acids present in the oil. Iodine value (IV) is a measure of unsaturation levels of the fatty acids. Cetane number (CN) provides an indication of ignition quality of the fuel. As the biodiesel properties of oil are a function of its fatty acid composition, it is important to understand the mechanism of regulation of fatty acid in oil (Sarin et al. 2007). Jatropha oil is a good substitute for petroleum derived diesel fuel as no or very little modification is required in the engine to use this as fuel (Agarwal and Agarwal 2007). Chhetri et al. (2008) found that Jatropha oil contains 45.79 per cent oleic acid, 32.27 per cent linoleic acid, 13.37 per cent palmitic acid and 5.43 per cent stearic acid. Diwani et al. (2009) determined the fatty acid composition and found linoleic acid (48.18 per cent) and oleic acid (28.46 per cent) as major components. Interestingly, Olaofe et al. (2010) found that linoleic acid was the prevalent fatty acid (34.6) and oleic acid was found to be very low (4.37 per cent). Rashid et al. (2010) found that Jatropha oil was comprised of linoleic (49.75 per cent), stearic (16.8 per cent), oleic (13 per cent), palmitic (12.15 per cent), arachidic (5.01 per cent) and gadoleic (2 per cent) acids. On the other hand, Gubitz et al. (1999) found a higher range of oleic acid (34.4-45.8 per cent) as compared to linoleic acid content (29-44.2 per cent). Daniel et al. (2011) reported very high oleic acid content (50.3 per cent) in *J. curcas*. Annarao et al. (2008) reported lipid content and profiling in Jatropha seeds at different developmental stages using H

NMR spectroscopy. The oil content increased from 0.3 to 24.40 per cent during seed development. The ratio of linoleic to linolenic acid increased at first 3 stages while only linoleic was observed at stage 4 onwards. These results suggested that linolenic acid was primarily involved in initial development at a stage when sterol concentration was high. It gave way to linoleic acid as the seeds started to mature.

Oil is present in the endosperm tissue of Jatropha seed. Both oil content and fatty acid composition are quantitative traits under polygenetic control which are greatly influenced by environmental conditions. Consequently, there has been interest in identifying QTLs controlling the lipid biosynthesis pathway. Fatty acid elongases, FAE1.1 and FAE1.2, have been found to be associated with oil content in *Brassica juncea* (Cheung et al. 1998; Gupta et al. 2004; Thormann et al. 1996) and *B. napus* (Ecke et al. 1995; Qiu et al. 2006).

5.7 Conclusion

During 2003-07, a large number of oil extraction and processing units had come up in different parts of India. Consequently, there was a great demand of Jatropha seeds to feed these processing units. It was soon felt that the key to success of the biodiesel initiative lied in the ability to provide the oil extraction / processing units a continuous supply of stable quality raw material i.e. seeds. This, in turn, was completely dependent on the availability of good quality planting material of *Jatropha*, capable of providing high seed yield with high oil content. However, there was a complete absence of any such kind of improved planting material in Jatropha. Most of the planting material available to growers was being sourced from untested spurious plantations. For improvement of Jatropha several studies were conducted and provide good information for its genetic modification so that it can be used as a future generation green fuel.

The overall aim of this research was to create a base for future genetic improvement of *J. curcas* in the form of genetic tools, information and breeding resources. The first prerequisite was to create a prebreeding material with enough genetic variability. Secondly, it was appreciated that developing various molecular markers and genomic tools such as linkage maps would potentially increase the efficiency of genetic improvement in this species. None of these genetic resources and tools was available for Jatropha. Researchers worked hard to raise the genetic diversity in Jatropha through interspecific hybridization. Molecular markers also played highly significant role to meet this purpose especially amplified fragment length polymorphism and microsatellite markers; former is dominant marker while later is co dominant. The greatest advantage of AFLP is that it can be used in any species without any prior nucleotide sequence information. For a newly studied species like Jatropha, little nucleotide sequence information was available at the time of this study. By using AFLP markers a core set of useful genotypes from the existing accessions were identified which could be used as parents in future breeding. Being oilseed crop, oil content was also considered during identification of core accessions. High oil containing accessions were selected for further hybridization purpose. Newly developed hybrids were further tested for oil content and proportion of useful fatty acids. High oil containing hybrids were sorted. Also different lipid profiles variants were also selected as per requirement for eg; high palmitic acid containing hybrids for soap production while high oleic acid containing hybrid for biodiesel production. These selected hybrids can be further utilized for fine mapping of Jatropha genome and can be released as variety.

One more aspect was focused in this chapter that is out crossing rates which is also very crucial for success of any plant breeding programme by conventional methods. Level of genetic diversity in a particular species is dependent on the mating system prevalent in the particular species. In general, out crossing rates are high in cross pollinated species while low in the self-pollinated species. Understanding of prevalent mode of pollination or mating system is important to develop a strategy for breeding and maintenance of populations. Although, Jatropha was thought to be an out crossed species due to its unisexual flowers, there were not much published studies available.

References

Adugna A, Sweeney PM and Bekele E (2013) Estimation of in situ mating systems in wild sorghum (*Sorghum bicolor* (L.) Moench) in Ethiopia using SSR-based progeny array data: implications for the spread of crop genes into the wild. Journal of Genetics 92: 3-10.

Agarwal D and Agarwal AK (2007) Performance and emissions characteristics of Jatropha oil (preheated and blends) in a direct injection compression ignition engine. Applied Thermal Engineering 27: 2314-2323.

Ahmed WA and Salimon J (2009) Phorbol ester as toxic constituents of tropical *Jatropha curcas* seed oil. European Journal of scientific research 31: 429-436.

Ahoton LE and Quenum F (2012) Floral biology and hybridization potential of nine accessions of Physic nut (*Jatropha curcas* L.) originating from three continents. Tropicultura 30: 193-198.

Allard RW, Jain SK and Workman PL (1968) The genetics of inbreeding populations. Advances in Genetics 14: 55-131.

Annarao S, Sidhu OP, Roy R, Tuli R and Khetrapal CL (2008) Lipid profiling of developing *Jatropha curcas* L. seeds using (1)H NMR spectroscopy. Bioresource Technology 99: 9032-9035.

Bamberg J and del Rio A (2014) Selection and validation of an AFLP marker core collection for the wild potato *Solanum microdontum*. American Journal of Potato Research 91: 368-375.

Banerji R, Chowdhury AR, Misra G, Sudarsanam G, Verma SC and Srivastava GS (1985) Jatropha seed oils for energy. Biomass 8: 277-282.

Barrett BA and Kidwell KK (1998) AFLP based genetic diversity assessment among wheat cultivars from the Pacific Northwest Crop Science 38: 1261-1271.

Barrett SCH and Kohn JR (1991) Genetic and evolutionary consequences of small population size in plants: implications for conservation. In: Falk DA, Holsinger KE (eds) Genetics and conservation of rare plants Oxford University Press, Oxford, pp 3-30.

Basha S and Sujatha M (2007) Inter and intra-population variability of *Jatropha curcas* (L.) characterized by RAPD and ISSR markers and development of population-specific SCAR markers. Euphytica 156: 375-386.

Basha SD, George F, Makkar HPS, Becker K and Sujatha M (2009) A comparative study of biochemical traits and molecular markers for assessment of genetic relationships between *Jatropha curcas* (L.) germplasm from different countries. Plant Science 176: 812-823.

Becker K and Makkar HPS (2008) *Jatropha curcas*: A potential source for tomorrow's oil and biodiesel. Lipid Technology 20: 104-107.

Bhattacharya A, Datta K and Datta SK (2005) Floral biology, floral resource constraints and pollination limitation in *Jatropha curcas* L. Pakistan Journal of Biological Sciences 8: 456-460.

Biabani A, Rafii MY, Saleh GB and Abdul LM (2013) Inter- and intra-population genetic variations in *Jatropha curcas* populations revealed by inter-simple sequence repeat molecular markers. Maydica 58: 111-118.

Biello D (2009) Green fuels for jets. Scientific American 19: 68-69.

Blum E, Liu K, Mazourek M, Yoo EY, Jahn M and Paran I (2002) Molecular mapping of the C locus for presence of pungency in Capsicum. Genome 45: 702-705.

Bos M, Steen R and Harmens H (1985) Protogyny in *Plantago lanceolata* populations: an adaptation to pollinate by wind? In: Jacquard P, Heim G, Antonovics J (eds) Genetic differentiation and dispersal in plants. Springer-Verlag, Berlin, pp 327-338.

Bressan EA, Scotton DC, Ferreira RR, Jorge EC, Sebbenn AM, Gerald LTS and Figueira A (2012) Development of microsatellite primers for *Jatropha curcas* (Euphorbiaceae) and transferability to congeners. American Journal of Botany 99: e237-e239.

Bressan EA, Sebbenn AM, Ferreira RR, Lee TSG and Figueira A (2013) *Jatropha curcas* L. (Euphorbiaceae) exhibits a mixed mating system, high correlated mating and apomixes. Tree Genetics & Genomics 9: 1089-1097.

Brown AHD (1979) Enzyme polymorphism in plant populations. Theoretical Population Biology 15: 1-42.

Brown AHD (1989) Core collections: a practical approach to genetic resources management. Genome 31: 818-824.

Carvalho CR, Clarindo WR, Praca MM, Araujo FS and Carels N (2008) Genome size, base composition and karyotype of *Jatropha curcas* L., an important biofuel plant. Plant Science 174: 613-617.

Chaix G, Gerber S, Razafimaharo V, Vigneron P, Verhaegen D and Hamon S (2003) Gene flow estimation with microsatellites in a Malagasy seed orchard of *Eucalyptus grandis*. Theoretical and Applied Genetics 107: 705-712.

Chang-Wei L, Kun L, You C and Young-Yu S (2007) Floral display and breeding system of *Jatropha curcas* L. Forest Studies China 10: 114-119.

Cheung WY, Landry BS, Raney P and Rakow GFW (1998) Molecular mapping of seed quality traits in *Brassica juncea* (L.) Czern and Coss. Acta Horticulturae 459: 139-147.

Chhetri A, Tango M, Budge S, Watts K and Islam M (2008) Non edible plant oils as new sources for biodiesel production. International Journal of Molecular Sciences 9: 169-180.

Collard BC and Mackill DJ (2008) Marker-assisted selection: an approach for precision plant breeding in the twenty-first century. Philosophical Transanctions of Royal Society of London B Biological Sciences 363: 557-572.

Cox TS, Murphy JP and Rodgers DM (1986) Changes in genetic diversity in the red winter wheat regions of the United States. Proceedings of National Academy of Sciences (USA) 83: 5583-5586.

Daniel L, Ardiyanti AR, Schuur B, Manurung R, Broekhuis AA and Heeres HJ (2011) Synthesis and properties of highly branched *Jatropha curcas* L. oil derivatives. European Journal of Lipid Science and Technology 113: 18-30

Dehgan B (1984) Phylogenetic significance of interspecific hybridization in Jatropha (Euphorbiaceae). Systematic Botany 9: 467- 478.

Dehgan B and Webster GL (1979) Morphology and infrageneric relationships of the genus Jatropha (Euphorbiaceae). University of California Publications in Botany, Vol. 74.

Desai CM and Vyas MT (1949) Chemical investigation of the seeds of *Jatropha curcas* Linn. Current Science 18: 49.

Dhillon RS, Hooda MS, Jattan M, Chawla V, Bhardwaj M and Goyal SC (2009) Development and molecular characterization of interspecific hybrids of *Jatropha curcas* x *Jatropha integerimma.* Indian Journal of Biotechnology 8: 384-390.

Diwani GEl, Attia NK and Hawash SI (2009) Development and evaluation of biodiesel fuel and by products from Jatropha oil. International Journal of Environmental Science and Technology 6: 219-224.

Ecke W, Uzunova M and Weissleder K (1995) Mapping the genome of rapeseed (*Brassica-napus* L.). 2. Localization of genes controlling erucic acid synthesis and seed oil content. Theoretical and Applied Genetics 91: 972-977.

Ennos RA (1985) The mating system and genetic structure in a perennial grass, *Cynosurus cristatus.* Heredity 55: 121-126.

Epperson BK and Clegg MT (1987) Frequency- dependent variation for outcrossing rate among flower color morphs of *Ipomoea purpurea.* Evolution 41: 1302-1311.

Frankel OH and Soule ME (1981) Conservation and evolution. Cambridge University Press, Cambridge.

Franklin JR (1980) Evolutionary changes in small populations. In: Soule ME, Wilcox BA (eds) Conservation biology: an evolutionary ecological perspective Sinauer Associates Sunderland, M A, pp 135-149.

Fresnedo-Ramirez J (2013) The floral biology of *Jatropha curcas* L. A review. Tropical Plant Biology 6: 1-15.

Gaiotto FA, Bramucci M and Grattapaglia D (1997) Estimation of outcrossing rate in a breeding population of *Eucalyptus urophylla* with dominant RAPD and AFLP markers. Theoretical and Applied Genetics 95: 842-849.

Ganesh Ram S, Parthiban K, Senthil Kumar R, Thiruvengadam V and Paramathma M (2008) Genetic diversity among Jatropha species as revealed by RAPD markers. Genetic Resources and Crop Evolution 55: 803-809.

Gubitz GM, Mittelbach M and Trabi M (1999) Exploitation of the tropical oil seed plant *Jatropha curcas* L. Bioresource Technology 67: 73-82.

Gunstone FD (2007) Research Highlights. Lipid Technology 19: 187-189.

Gupta V, Mukhopadhyay A, Arumugam N, Sodhi YS, Pental D and Pradhan AK (2004) Molecular tagging of erucic acid trait in oilseed mustard (*Brassica juncea*) by QTL mapping and single nucleotide polymorphisms in *FAE1* gene. Theoretical and Applied Genetics 108: 743-749.

Gusson E, Sebbenn AM and Kageyama PY (2006) Sistema de reproducao em populacoes de *Eschweilera ovata* (Cambess.) Miers. Revista Árvore 30: 491-502.

Hallauer AR and Miranda JB (1988) Quantitative genetics in maize breeding, 2nd edition Iowa state University Press, Ames, IA.

Hamrick JL and Godt MJW (1989) Allozyme diversity in plant species. In: Brown AHD, Clegg MT, Kahler AL, Weir BS (eds) Conservation genetics: case histories from natureChapman and Hall, New York, pp 281-304.

He W, King AJ, Khan MA, Cuevas JA, Ramiaramanana D and Graham IA (2011) Analysis of seed phorbol-ester and curcin content together with genetic diversity in multiple provenances of *Jatropha curcas* L. from Madagascar and Mexico Plant Physiology and Biochemistry 49: 1183-1190.

Heller J (1996) Physic nut. *Jatropha curcas* L. Promoting the conservation and use of underutilized and neglected crops Institute of Plant Genetics and Crop Plant Research, Gatersleben/ International Plant Genetic Resources Institute, Rome.

Hirakawa H, Tsuchimoto S, Sakai H et al (2012) Upgraded genomic information of *Jatropha curcas L.* Plant Biotechnology Journal 29: 123-130.

Jansen J and van Hintum T (2007) Genetic distance sampling: a novel sampling method for obtaining core collections using genetic distances with an application to cultivated lettuce. Theoretical and Applied Genetics 114: 421-428.

Karmee SK and Chadha A (2005) Preparation of biodiesel from crude oil of *Pongamia pinnata.* Bioresource Technology 96: 1425-1429.

Kaur K, Dhillon GPS and Gill RIS (2011) Floral biology and breeding system of *Jatropha curcas* in north-western India. Journal ofTropical Forest Science 23: 4-9.

Kaushik N, Kumar K, Kumar S, Kaushik N and Roy S (2007) Genetic variability and divergence studies in seed traits and oil content of Jatropha (*Jatropha curcas* L.) accessions. Biomass and Bioenergy 31: 497-502.

Kesseli RV and Jain SK (1985) Breeding systems and population structure in Limnanthes. Theoretical and Applied Genetics 71: 292-299.

King AJ, Montes LR, Clarke JG et al (2013) Linkage mapping in the oilseed crop *Jatropha curcas* L. reveals a locus controlling the biosynthesis of phorbol esters which cause seed toxicity. Plant Biotechnology Journal 11: 986-996.

Kumar RS, Parthiban KT, Hemalatha P, Kalaiselvi T and Rao MG (2009) Investigation on cross-compatibility barriers in the biofuel crop *Jatropha curcas* L. with wild Jatropha species. Crop Science 49: 1667-1674.

Kywe TT and Oo MM (2009) Production of biodiesel from Jatropha oil (*Jatropha curcas*) in pilot plant. Proceedings of World Academy of Science, Engineering & Technology 38: 481-487.

Laosatit K, Tanya P, Muakrong N and Srinives P (2014) Development of interspecific and intergeneric hybrids among jatropha related species and verification of the hybrids using EST-SSR markers Plant Genetic Resources 12: S58-S61.

Levin DA and Kerster H (1969) Density-dependent gene dispersal in Liatris. The American Naturalist 103: 61-74.

Levin DA and Kerster HW (1974) Gene flow in seed plants. Evolutionary Biology 7: 139-220.

Liu P, Wang CM, Li L, Sun F and Yue GH (2011) Mapping QTLs for oil traits and eQTLs for oleosin genes in Jatropha. BMC Plant Biology 11: 132.

Luo CW, Huang ZY, Chen XM, Li K, Chen Y and Sun YY (2011) Contribution of diurnal and nocturnal insects to the pollination of *Jatropha curcas* (Euphorbiaceae) in southwestern China. Journal of Economic Entomology 104: 149-154.

Luo CW, Li K, Chen XM and Huang ZY (2012) Ants contribute significantly to the pollination of a biodiesel plant, *Jatropha curcas*. Environmental Entomology 41: 1163-1168.

Marshall DF and Abbott RJ (1982) Polymorphism for outcrossing frequency at the ray floret locus in *Senecio vulgaris* L. I. Evidence. Heredity 48: 227-235.

Martin GB, Brommonshenkel S, Chunwongse J et al (1993) Map-based cloning of a protein kinase gene conferring disease resistance in tomato. Science 262: 1432-1435.

Mohammadi SA and Prasanna BM (2003) Analysis of genetic diversity in crop plants: Salient statistical tools and considerations. Crop Science 43: 1235-1248.

Muller HJ (1932) Some genetic aspects of sex. American naturalists 66: 118-138.

Muluvi GM, Sprent JI, Odee D and Powell W (2004) Estimates of outcrossing rates in *Moringa oleifera* using amplified fragment length polymorphism (AFLP). African Journal of Biotechnology 3: 146-151.

Na-ek Y, Wongkaew A, Phumichai T, Kongsiri N, Kaveeta R, Reewongchai T and Phumichai C (2011) Genetic diversity of physic nut (*Jatropha curcas*

L.) revealed by SSR markers. Journal of Crop Science and Biotechnology 14: 105-110.

Naik M, Meher LC, Naik SN and Das LM (2008) Production of biodiesel from high free fatty acid Karanja (*Pongamia pinnata*) oil. Biomass and Bioenergy 32: 354-357.

Olaofe O, faleye FJ and Abey JA (2010) Fatty acid and physic-chemical analyses of *Jatropha curcas* (Physic nut) seed flour and oil. Bangladesh Journal of Science and Industrial Research 45: 345-350.

Osorio LRM, Salvador AFT, Jongschaap REE et al (2014) High level of molecular and phenotypic biodiversity in *Jatropha curcas* from Central America compared to Africa, Asia and South America. BMC Plant Biology 14: 77.

Pamidimarri DVNS, Chattopadhyay B and Reddy MP (2009a) Genetic divergence and phylogenetic analysis of genus Jatropha based on nuclear ribosomal DNA ITS sequence. Molecular Biology Reports 36: 1929-1935.

Pamidimarri DVNS, Mastan SG, Rahman H, Prakash Ch R, Singh S and Reddy MP (2011) Cross species amplification ability of novel microsatellites isolated from *Jatropha curcas* and genetic relationship with sister taxa: cross species amplification and genetic relationship of Jatropha using novel microsatellites. Molecular Biology Reports 38: 1383-1388.

Pamidimarri DVNS, Pandya N, Reddy MP and Radhakrishnan T (2009b) Comparative study of interspecific genetic divergence and phylogenic analysis of genus Jatropha by RAPD and AFLP. Molecular Biology Reports 36: 901-907.

Pamidimarri DVNS, Rahman H, Mastan SG and Reddy MP (2010) Isolation of novel microsatellites using FIASCO by dual probe enrichment from *Jatropha curcas* L. and study on genetic equilibrium and diversity of Indian population revealed by isolated microsatellites. Molecular Biology Reports 37: 3785-3793.

Pamidimarri DVNS, Sinha R, Kothari P and Reddy MP (2009c) Isolation of novel microsatellites from *Jatropha curcas* L. and their crossspecies amplification. Molecular Ecology Resources 9: 431-433.

Parthiban KT, Kumar RS, Thiyagarajan P, Subbulakshmi V, Vennila S and Rao MG (2009) Hybrid progenies in Jatropha– a new development. Current Science 96: 815-823.

Phumichai C, Phumichai T, Kongsiri N, Wongkaew A, Sripichit P and Kaveeta R (2011) Isolation of 55 microsatellite markers for *Jatropha curcas* and its closely related species. Biologia Plantarum 55: 387-390.

Pollard M and Ohlrogge J (1999) Testing models of fatty acid transfer and lipid synthesis in spinach leaf using in vivo oxygen- 18 labelling. Plant Physiology 121: 1217-1226.

Popluechai S, Breviario D, Mulpuri S, Makkar H, Raorane M, Reddy A, Palchetti E, Gatehouse A, Syers K, O'Donnell A and Kohli A (2009) Narrow genetic and apparent phenetic diversity in *Jatropha curcas*: initial success with generating low phorbol ester interspecific hybrids. Available from Nature Precedings <http://hdl.handle.net/10101/npre.2009.2782.1.

Prabakaran AJ and Sujatha M (1999) *Jatropha tanjorensis* Ellis & Saroja, a natural interspecific hybrid occurring in Tamil Nadu, India. Genetic Resources and Crop Evolution 46: 213-218.

Pranesh KJ, Rao MJ, Soumya HC, Gowda B, Savithramma and Naveen NL (2010) Studies on floral display and mode of reproduction in Jatropha (*Jatropha curcas* L.). Electronic Journal of Plant Breeding 1: 832-838.

Qiu D, Morgan C, Shi J et al (2006) A comparative linkage map of oilseed rape and its use for QTL analysis of seed oil and erucic acid content. Theoretical and Applied Genetics 114: 67-80.

Raju AJS and Ezradanam V (2002) Pollination ecology and fruiting behaviour in a monoecious species, *Jatropha curcas* L. (Euphorbiaceae). Current Science 83: 1395-1398.

Rashid U, Anwar F, Jamil A and Bhatti HN (2010) *Jatropha curcas* seed oil as a viable source for biodiesel. Pakistan Journal of Botany 42 575-582.

Ritland K (2002) Extensions of models for the estimation of mating systems using n independent loci. Heredity 88: 221-228.

Ritland K and Jain SA (1981) Model for the estimation of outcrossing rate and gene frequencies using "n" independent loci. Heredity 47: 35-52.

Rosado TB, Laviola BG, Faria DA et al (2010) Molecular markers reveal limited genetic diversity in a large germplasm collection of the biofuel crop *Jatropha curcas* L. in Brazil. Crop Science 50: 2372-2382.

Rupert EA, Dehgan B and Webster GL (1970) Experimental studies of relationships in the genus Jatropha L. *J. curcas* x *J. integerimma.* Bulletin of Torrey Botanical Club 99: 321-325.

Santos CAF and Gama RNCdS (2013) An AFLP estimation of the outcrossing rate of *Spondias tuberose* (Anacardiaceae), an endemic species to the Brazilian semiarid region. Rev. Biol. Trop 61: 577-582.

Sarin R, Sharma M, Sinharay S and Malhotra RK (2007) Jatropha–palm biodiesel blends: an optimum mix for Asia. Fuel 86: 1365-1371.

Sato S, Hirakawa H, Isobe S et al (2010) Sequence analysis of the genome of an oil-bearing tree, *Jatropha curcas* L. DNA Research 18: 65-76.

Schmitt J (1983) Density-dependent pollinator foraging, flowering phenology and temporal pollen dispersal patterns in *Linanthus bicolor*. Evolution 37: 361-370.

Schoen DJ (1982) The breeding system of *Gilia achilleifolia*: variation in floral characteristics and outcrossing rate. Evolution 36: 352-360.

Schoen DJ and Brown AHD (1993) Conservation of allelic richness in wild crop relatives is aided by assessment of genetic markers. Proceedings of the National Academy of Sciences USA 90: 10623-10627.

Schonewald- Cox CS, Chambers SM, Macbryde B and Thomas L (1983) Genetics and conservation: a reference for managing wild animal and plant populations Benjamin- Cummings, London.

Sharma A and Chauhan RS (2011) Repertoire of SSRs in the castor bean genome and their utilization in genetic diversity analysis in *Jatropha curcas*. Computational and Functional Genomics 28: 60-89.

Sinha P, Dwivedi N, Negi MS and Tripathi SB (2014) Genetic variability among Jatropha species as revealed by amplified fragment length polymorphism (AFLP) markers. Indian Journal of Biotechnology 13: 496-501.

Sinha P, Md Islam A, Negi MS and Tripathi SB (2015b) Development of novel microsatellite markers in *Jatropha curcas* and evaluation of their cross-species transferability. Proceedings of the National Academy of Sciences, India Section B: Biological Sciences. .

https://doi.org/10.1007/s40011-40015-40489-x

Sinha P, Md Islam A, Negi MS and Tripathi SB (2015c) First identification of core accessions of *Jatropha curcas* from India based on molecular genetic diversity. Plant Genetic Resources. https://doi.org/10.1017/S1479262115 000088.

Sinha P, Negi MS, Sharma SS, Md Islam A and Tripathi SB (2013) Analysis of genome-wide homozygosity in *Jatropha curcas* accessions using AFLP markers. International Journal of Research in Pharmacy and Science 3: 191-201.

Sinha P and Tripathi SB (2013) Molecular characterization of Jatropha genetic resources through amplified fragment length polymorphism (AFLP) markers. International Journal of ChemTech Research 5: 735-740.

Smith JSC (1984) Genetic variability within US hybrid maize: Multivariate analysis of isozyme data. Crop Science 24: 1041-1046.

Stalker HT (1980) Utilization of wild species for crop improvement. Advances in Agronomy 33: 111-147.

Stern K and Roche F (1974) Genetics of forest ecosystems. Springer Verlag, Berlin.

Subramanyam K, Muralidhararao D and Devanna N (2009) Genetic diversity assessment of wild and cultivated varieties of *Jatropha curcas* (L.) in India by RAPD analysis. African Journal of Biotechnology 8: 1900-1910.

Sujatha M and Prabakaran AJ (2003) New ornamental Jatropha hybrids through interspecific hybridization. Genetic Resources and Crop Evolution 50: 75-82.

Sun F, Liu P, Ye J, Lo LC, Cao S, Li L, Yue GH and Wang CM (2012) An approach for Jatropha improvement using pleiotropic QTLs regulating plant growth and seed yield. Biotechnology for Biofuels 5: 42.

Sun Q-B, Li L-F, Li Y, Wu G-J and Ge X-J (2008) SSR and AFLP markers reveal low genetic diversity in the biofuel plant *Jatropha curcas* in China. Crop Science 48: 1865-1871.

Suwannoi S, Kanchanaketu T, Kusolkumbot P, Sangduen N and Hongtrakul V (2011) DNA methylation and genetic study of interspecific hybridization in *Jatropha curcas* L. Thai Journal of Genetics 4: 94-105.

Tanya P, Dachapak S, Tar MM and Srinives P (2011) New microsatellite markers classifying nontoxic and toxic *Jatropha curcas*. Journal of Genetics 90: e76-e78.

Tatikonda L, Wani SP, Kannan S, Beerelli N, Sreedevi TK, Hoisington DA, Devi P and Varshney RK (2009) AFLP-based molecular characterization of an elite germplasm collection of *Jatropha curcas* L., a biofuel plant. Plant Science 176: 505-513.

Thompson JA, Nelson RL and Vodkin LO (1998) Identification of diverse soybean germplasm using RAPD markers. Crop Science 38: 1348-1355.

Thormann CE, Romero J, Mantet J and Osborn TC (1996) Mapping loci controlling the concentrations of erucic and linolenic acids in seed oil of *Brassica napus* L. Theoretical and Applied Genetics 93: 282-286.

Wang CM, Liu P, Yi C et al (2011) A first generation microsatellite- and SNP-based linkage map of Jatropha. PLoS ONE 6: e23632.

Wang Z, Zhuge J, Fang H and Prior BA (2001) Glycerol production by microbial fermentation: A review. Biotechnology Advances 19: 201-223.

Warrier RR, Nagalakshmi SC, Anandalakshmi R, Nicodemus A and Singh GB (2012) Assessment of macro geographical genetic variations in *Jatropha curcas* L. in India using allozyme and RAPD markers. Tropical Agricultural Research and Extension 15.

Watkins L and Levin DA (1990) Outcrossing rates as related to plant density in *Phlox drummondii*. Heredity 65: 81-89.

Wen M, Wang H, Xia Z, Zou M, Lu C and Wang W (2010) Development of EST-SSR and genomic-SSR markers to assess genetic diversity in *Jatropha curcas* L. BMC Research Notes 3: 42.

Wolff K, Friso B and van Damme JMM (1988) Outcrossing rates and male sterility in natural populations of *Plantago coronopus*. Theoretical and Applied Genetics 76: 190-196.

Wu P, Zhou C, Cheng S et al (2015) Integrated genome sequence and linkage map of physic nut (*Jatropha curcas* L.), a biodiesel plant. The Plant Journal 81: 810-821.

Yadav HK, Ranjan A, Hasan AM, Mantri S, Sawant S and Tuli R (2011) EST-derived SSR markers in *Jatropha curcas* L.: development, characterization, polymorphism, and transferability across the species/genera. Tree Genetics & Genomes 7: 207-219.

Yi G, Lee JM, Lee S, Choi D and Kim B-D (2006a) Exploitation of pepper EST–SSRs and an SSR-based linkage map. Theoretical and Applied Genetics 114: 113-130.

Zhang Z, Guo X, Liu B, Tang L and Chen F (2011) Genetic diversity and genetic relationship of *Jatropha curcas* between China and Southeast Asian revealed by amplified fragment length polymorphisms. African Journal of Biotechnology 10: 2825-2832.

Chapter - 6

Junk DNA: Its Role in Plant Gene Regulation and Crop Improvement

—Santosh Kumar and Sabhyata Bhatia

6.1 Introduction

Microsatellites (Litt and Luty, 1989), also known as short tandem repeats (STRs), simple sequence repeats (SSRs) are tandem repeat motifs of 1-6 bp which have a frequent occurrence in all prokaryotic and eukaryotic genomes (Zane et al. 2002). Such DNA elements which were sometimes termed as "junk DNA" can serve multiple purposes in the genomes of higher eukaryotes. Microsatellites and their function in higher genomes is an interesting subject of contemporary research. While some investigators considered microsatellites as junk DNA, being too simple in structure to have a functional role, others suggested that simplicity itself may serve many functional roles. In plants, despite extensive reports of SSR identification and distribution within ESTs, their role in gene regulation remains to be investigated (Li et al. 2004). There is accumulating evidence that SSRs are non-randomly distributed across protein-coding regions, UTRs and introns (Fujimori et al. 2003) and may be involved in various potential functions in regulation of gene expression. Often, these elements have been found to be associated with gene regulation (Simpson and Ayyar, 2008), functional evolution of genes (Biemont and Vieiars 2006) and genome evolution. Moreover, it has been recently shown that SSRs are more abundant in the untranslated regions (UTRs) or regulatory regions than in other genomic regions of plants (Morgante et al. 2002). The location of SSRs in the genome determines its functional role and may have the potential to affect all aspects of genetic functions including gene regulation, development and evolution. A microsatellite located in a coding region can affect the activation of a gene and therefore, the expression of a protein. If located in a noncoding region e.g., the 5'-untranslated regions (UTRs) or introns, the microsatellite may impact gene transcription (Lawson and Zhang, 2006). The microsatellite sequences within 5' UTR region or downstream of transcription start site of a gene are very crucial as not only they have a role in protein translation but recently several reports have demonstrated that they help in promoter activity and may be the members of core promoter elements. It is now apparent that the core promoter is a structurally and functionally diverse transcriptional regulatory element. They can be viewed at the level of DNA sequences, transcriptional factors, and biological networks. There are no universal core promoter elements that are present in all promoters. A recently described class of elements, located downstream of the classical core promoter elements, has added another level of complexity to the core promoter architecture. These include the downstream promoter element (DPE) (Burke et al., 1997), the downstream core element (DCE) (Anish et al. 2009) and motif ten element (MTE) (Lim et al. 2004). Therefore, sequences (novel) in and around the transcription start site may have the potential to become a member of core promoter elements. In plants, the preferential distribution of microsatellites in 5'-UTRs (Fujimori et al. 2003; Morgante et al. 2002), and a novel plant specific feature in which a gradient in the microsatellite density along the direction of transcription (Fujimori et al. 2003), makes them a strong contender to become a regulatory element.

The SSRs /Junk DNA may act as "tuning knobs" (Kashi et al. 1997; Trifonov, 2003) for modulation of gene expression under various environmental challenges that cause various stress reactions, including changes in the copy number of tandem repeats. This observation is supported by the fact that DNA damage caused by external stresses such as UV irradiation, t-butyl hydrogen peroxide, oxidative damage, etc. can induce slippage mutations and increase mutation rates in SSR sequences (Jackson et al. 1998). An immediate response to environmental challenges is retuning of the expression of many genes and multigene functions may be influenced by the repeats. The copy numbers of the variable "tuners" linked to specific genes most relevant to given environmental changes are under selection pressure. Larger the number of repeats in the locus, the finer the tuning. Accordingly, some of the copy-number tuning responses to the stress result in observable phenotypic changes. The tuning functions of SSRs for a number of genes in eukaryotes and prokaryotes, suggest that tuning by the repeat copy number is a rather general phenomenon. SSRs are overrepresented among ORFs encoding for regulatory proteins (e.g., transcription factors and protein kinases) rather than for structural ones, indicating the role of SSRs as a factor contributing to fast evolution of adaptive changes (Young et al. 2000). Moreover, recently, many reports have demonstrated that a large number of SSRs are located in transcribed regions of genomes, including protein-coding genes and expressed sequence tags (ESTs) (Morgante et al. 2002). These ESTs have a range of functions such as metabolic enzymes, structural and storage proteins, disease signaling, and transcription factors, suggesting some role(s) of SSRs in plant metabolism and gene evolution.

Catharanthus roseus (Madagascar periwinkle) is widely known for its pharmacologically important alkaloids such as the anti-cancer compounds vinblastine and vincristine. The plant exhibits an unsurpassed spectrum of chemodiversity as it produces over 130 alkaloids through the terpenoid indole alkaloid (TIA) pathway (van der Heijden et al. 2004). All the terpenoid indole alkaloids (TIA) are derived from the precursor strictosidine, which is formed by condensation of indole tryptamine and the terpenoid secologanin in a reaction catalyzed by the enzyme strictosidine synthase. However, tryptophan decarboxylase (*Tdc*) converts tryptophan into tryptamine, and is a key enzyme in the biosynthetic pathway due to its position on the interface of primary and secondary metabolism. These two key TIA pathway genes strongly respond to a group of chemical or fungal elicitors via transcription factors. Moreover, these two genes have been shown to have SSRs in their 5'- UTRs, but functional role of these SSRs remains to be discovered. The immense potential of the TIA pathway is yet to be unravelled, generating a renewed interest in various aspects of *C. roseus* research.

6.2 Function of SSRs in genes

For quite a long time, microsatellites were only considered as genetic markers in DNA fingerprinting and diversity studies due to the extensive length polymorphisms. There are few reports of functional feature of microsatellites in the transcribed regions of plant (Table 6.1) and their location in the genome determines its functional role (Fig. 6.1). These have the potential to affect all aspects of genetic functions including gene regulation, development and evolution. A microsatellite located in a coding region can affect the activation of a gene and therefore, the expression of a protein. If located in a noncoding or genic region, e.g., the 5'-untranslated regions (UTRs) or introns, the microsatellite may impact gene regulation or gene transcription (Lawson and Zhang, 2006). Presence of SSRs in the 5'-UTRs are required for expression of some genes and variations in SSRs could regulate expression of these genes by affecting transcription and translation. For example, a polymorphic pentanucleotide repeat located within the 5'UTR of the p53-induced gene 3 (*PIG3*) and the repeat is necessary and sufficient for transcriptional activation of *PIG3* by p53 (Contente et al. 2002). The SSR expansions in 3'UTRs cause transcription slippage and produce expanded mRNA, which can disrupt splicing and other cellular functions. Intronic SSRs can also regulate gene transcription and translation. For example, different alleles of tetranucleotide repeat (TACT) located in the first intron of tyrosine hydroxylase gene (*TH*) were shown to differ in their silencing effect on *TH* expression (Albanese et al. 2001) In addition, it can also behave as a co-regulator for gene expression alongwith SSRs in the 5'-UTR. SSR expansion in introns and 3'-UTRs can lead to gene silencing or a loss of function. All these effects caused by SSR expansion or contraction within gene can ultimately lead to phenotypic changes. These unusual phenotypic variations cause several diseases in humans (Li et al. 2004). SSR expansion and/or contraction in exon can lead to a gain or loss of gene function via frame shift mutation or expanded mRNA toxicity (Li et al. 2004). It had been reported that microsatellites such as (CT)*n*/(GA)*n* and (CTT)*n*/(GAA)*n* found in regulatory regions of many genes can affect chromatin structure and DNA conformation (Toth et al. 2000).

Table 6.1: Role of microsatellites in plants

Type	*Motifs Name*	*Recognized Sequence*	*Motifs Function*
(GA)n	GAGAGMGSA	$(GA)_9$	Binding site for GAGA- binding factor, and Gbp is a light-responsive gene (Sangwan and O'Berian 2002).
(CT)n	TCCC-motif	TCTCtCT	Part of light responsive element (Bolle et al. 1996).
	CTRMCAMV35S	TCTCTCTCT	CT-rich motif found in 60- nt region downstream of TSS of CaM 35S RNA; can enhance gene expression (Pauli et al. 2004).
	CT/GA	…………	Variation in the number of CT/GA repeats in the 5'UTR of the waxy gene, corre lated with anylose content in rice (Bao et al. 2002).
(CT)n	CT/AG- element		DPE like element (Kumar et al. 2016)

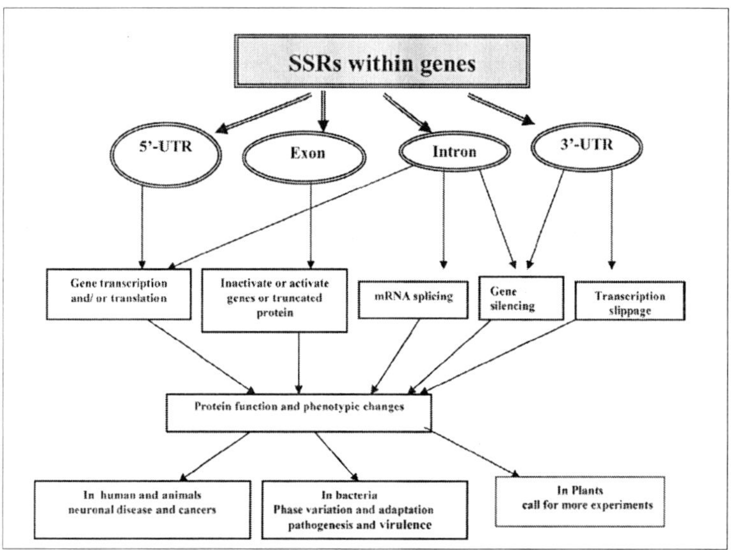

Figure 6.1: ***SSR regulatory functions in exon-intron and UTRs (5' and 3') adapted from Li et al. (2004).***

6.3 'Junk DNA' Topology and transcription

These repetitive sequences can adopt non B-DNA structure called H-DNA. Since DNA undergoes a characteristic conformational change upon H-DNA formation and forms a sharply bent double helix, which is often observed when DNA- regulatory proteins bind to DNA. Zain et al. (2003) have postulated that H-DNA could act as a molecular switch to modulate gene expression in a structure-dependent manner *in vivo*, and that cellular protein could specifically recognize triple-helical DNA structures and stabilize them. This speculation was supported by two facts: (1) Mirror repeats polypurine-polypyrimidine sequence have been found in several eukaryotic genome, and located in particular near regulatory regions (2) DNA topology constraints, especially negative supercoiling, are an intrinsic and dynamic feature during the processing of DNA information such as replication and transcription (Zain et al. 2003). Indeed, (CT)*n*.(AG)*n* dinucleotide sequences function as a positive transcription element to stimulate gene expression (Lu et al. 2003). Moreover, the *trans*-acting GAGA factor has been shown to bind dinucleotide repeats and to stimulate promoter activities of heat shock genes in *Drosophila melanogaster* by establishing nucleosome-free DNase-hypersensitive sites (Lu et al. 2003). Moreover, the extended CT.AG repeats in the promoter increases H-DNA structures, histone modification, and promoter activity. However, the function of genes that contain SSRs and the role of these SSR motifs in plant genes are less studied and poorly understood (Kumar et al. 2016). Fujimori et al. (2003) suggested that microsatellites located at high frequency in the 5'- flanking regions of plant genes can potentially act as factors in regulating gene expression. Some of the microsatellites may act as *cis*-regulatory elements which can be recognized by transcription factors (Martin et al. 2004). It has been well known for so called GAGA elements, comprising the dinucleotide repeat sequence (GA)*n* to be present in promoters regulating numerous developmental genes in animals (Busturia et al. 2001). Similarly, the (GA)*n* sequences in regulatory regions of some plant genes can also be recognized by GAGA-binding factors (Sangwan et al. 2002; Santi et al. 2003), and more generally, the GA-rich element, a more complex 9 base pairs (bp) based (GA)*n* repeat, has been shown to have protein-binding affinity (Kooiker et al. 2005). Another major microsatellite in plants, the trinucleotide repeat sequence (GAA)*n* presented within 5'-UTR of *ntp303* was found important in the modulation of transcription and translation efficiency (Hulzink et al. 2002). Furthermore, some unusual phenotypic variations were found to be associated with the length of 5'-noncoding microsatellites. A typical example was reported by Bao and his colleagues that variation in the number of CT/GA repeats in the 5'-UTR of the *waxy* gene was correlated with amylose content in rice (Bao et al. 2000). Although the mechanism is still unclear, the microsatellite length polymorphism is thought to affect the expression of the related genes of amylose synthesis. A recent study revealed that microsatellites such as (CT)n/(GA)*n* and (CTT)*n*/(GAA)*n* are well presented in plant genome and in particular, they are preferentially located within the 5'-noncoding regions (Kumar et al. 2016). These 5'-noncoding microsatellites, served as regulatory element involved in light and salicylic acid response (Zhang et al. 2006).

Nucleic acids are polymorphic macromolecules and can adopt a variety of single-stranded, duplex and multistranded conformations. The potential importance of homopurine-homopyrimidine sequences in biological processes is suggested by the ubiquity and non-random distribution of sequences such as (CT)n.(GA)*n* within the genomes of many organisms (Toth et al. 2000). The (CT)n.(GA)*n* repeats have been found in promoter regions of many genes, at recombination hotspots and at replication origins (Wells et al. 1998). Deletion or mutation of these sequences in promoter regions reduces gene expression, implying that these sequences have functional roles (Glaser et al. 1990). These sequences are unique in that they form noncanonical DNA secondary structures at promoter regions of many genes. Along with the trinucleotide (GAA)*n*.(TTC)*n* or (CTT)*n*. (AAG)*n* repeats, dinucleotide (CT)*n*.(AG)*n* sequences can adopt a non-B DNA structure, an intramolecular triplex or H-DNA. These structures form at regions containing homopurine-homopyrimidine sequences, where part of the tract can dissociate into single strands using the energy provided by supercoiling. One of the single strands can then swivel its backbone parallel to the duplex, forming a three-stranded helix and leaving its complementary strand unpaired (Htun and Dehlberg, 1989). Since DNA undergoes a characteristic conformational change upon H-DNA formation and forms a sharply bent double helix, which is often observed when DNA- regulatory proteins bind to DNA. Zain and Sun, (2003) have postulated that H-DNA could act as a molecular switch to modulate gene expression in a structure- dependent manner *in vivo*, and that cellular protein could specifically recognize triple-helical DNA structures and stabilize them. This speculation was supported by two facts: (1) Mirror repeats polypurine-polypyrimidine sequence have been found in several eukaryotic genomes, and located in particular near regulatory regions (2) DNA topology constraints, especially negative supercoiling, are an intrinsic and dynamic feature during the processing of DNA information such as replication and transcription (Zain et al. 2003). Indeed, (CT)*n*.(AG)*n* dinucleotide sequences function as a positive transcription element to stimulate gene expression (Lu et al. 1993), and deletion of these sequences reduced promoter activity of heat shock genes in *Drosophila melanogaster* (Sandaltzopoulos et al. 1995). Moreover, the *trans*-acting GAGA factor has been shown to bind dinucleotide repeats and to stimulate promoter activities of heat shock genes in *Drosophila melanogaster* by establishing nucleosome-free DNase-hypersensitive sites (Lu et al. 2003). Moreover, the extended CT.AG repeats in the promoter increases H-DNA structures, histone modification, and promoter activity (Han and de Lanerolle, 2008). Nevertheless, *in vivo* establishment of a detailed mechanism of an intramolecular triplex-directed regulation of gene expression is still very challenging.

6.4 Summary

Microsatellites, as one of the major repeat classes, are extremely common in eukaryotic genomes. They are generally thought to result from the mutation effects of replication slippage. Plant microsatellites show a significant association with non-repetitive DNA. They can be found abundantly within or near genes in plant genomes and in particular, some types are significantly enriched within the 5' non-coding regions of plant genes. For example, in *Arabidopsis thaliana*, this feature is mostly attributed to the fact that CT/GA and CTT/GAA repeats are more frequently found in 5'-flanks than in other genomic regions. Such DNA elements which were sometimes termed as "junk DNA" can serve multiple purposes in the genomes of higher eukaryotes. For quite a long time, microsatellites were only considered as genetic markers in DNA fingerprinting and diversity studies due to the extensive length polymorphisms. However, recent findings show that some of them act as *cis*-regulatory elements which can be recognized by transcription factors. The effect of long di- and tri-nucleotide SSRs on the transcript level depends on the position of the SSRs within the gene regions. While the presence of long dinucleotide repeats within introns and 3' regions was often associated with lowering the level of expression, genes with such repeats in 5' regions are associated with increasing the transcript levels (Sharopova, 2008). It has been well known for so called GAGA elements, comprising the dinucleotide repeat sequence (GA)n, to be present in promoters regulating numerous developmental genes in animals. Similarly, the (GA)n sequences in regulatory regions of some plant genes can also be recognized by GAGA-binding factors and more generally, the GA-rich element, a more complex 9 base pairs (bp) based (GA)n repeat, has been shown to have protein-binding affinity. Another major microsatellite in plants, the trinucleotide repeats sequence (GAA)n present within 5'UTR of *ntp303* was found important in the modulation of transcription and translation efficiency. Another example was reported by Bao et al. 2002 that variation in the number of CT/GA repeats in the 5'UTR of the *waxy* gene was correlated with amylose content in rice. Although the mechanism is still unclear, the microsatellite length polymorphism is thought to affect the expression of the related genes of amylose synthesis.

A detailed analysis of the *C. roseus* transcriptome revealed that the frequency distribution of repeat motifs differs between the various fractions of the genic regions. Comparison of microsatellite occurrences showed that the CT/AG and CTT/AAG repeats were preferentially associated with 5'-flanking region of genes and the presence of these repeats may possibly modulate their expression in *C. roseus*. Such types of microsatellites have been found to be positively selected in regions downstream of the TSS and have been shown to be highly conserved in the housekeeping genes of plants and perhaps have a significant functional role. The tryptophan decarboxylase (*Tdc*) gene of the TIA pathway which is known to harbor the (CT)*n* repeat motifs in the proximity of the TSS in the 5' UTR was selected. The *C. roseus* accessions were analysed for length polymorphism in the 5' UTR of the *Tdc* gene. Sequence comparison of the *Tdc* promoter regions from Kew1 and Prabal also revealed that the only sequence variation that existed was in the copy number of the microsatellite motifs. In order to investigate the effects of 5'UTR SSRs on *Tdc* gene expression, a series of experiments and comparative analysis was performed, which included investigating 1) the effect of microsatellite length polymorphism present in the 5'-UTRs of *Tdc* gene on level of *Tdc* expression 2) role of length polymorphism in environmental stress mitigation 3) effects of length polymorphism on rate of *Tdc* gene transcription and 4) analysis of protein- DNA (microsatellite) interaction.

Expression analysis revealed that a positive correlation existed between length of microsatellite repeats and mRNA transcript abundance of *Tdc* gene in *C. roseus* accessions. Furthermore, it was observed that not only the length of dinucleotide SSRs influences the level of gene expression but the presence of SSRs *per se* in the 5'-UTR significantly enhances *Tdc* gene expression. This was evident from the fact that both natural variation in the CT/GA repeats in the 5' UTR of *Tdc* gene and/or in chimeric construct, harboring same repeats upstream of reporter gene (*gusA*) in transient assay, led to enhanced expression of respective gene in a copy number dependent manner. Furthermore, an immediate response to environmental challenges by 'tuning' the expression of *Tdc* gene influenced by the number of repeats was observed wherein longer repeats strongly responded to environmental stress such as UV & MeJA, when compared to shorter repeats. Thus 5'UTR SSRs may act as "tuning knob" for modulation of *Tdc* gene expression, where larger the number of repeats in tandem, finer the tuning of *Tdc* gene transcription. Further, it was observed that the rate of *Tdc* gene transcription was affected by microsatellite length polymorphism. This was confirmed by the nuclear run-on assay that established that mRNA transcript abundance was due to enhanced rate of *Tdc* gene transcription and not due to the role of repeats in mRNA stability. Furthermore, a *trans*-acting protein factor was shown to interact with single stranded conformation of microsatellite DNA *in-vitro*, thereby suggesting that the microsatellite sequence flanking the transcription start site might be a new *cis*-acting down-stream promoter element (DPE) and the binding of proteins to tandemly repeated sequences may bring about a higher level of *Tdc* gene transcription in a length dependent fashion. This phenomenon was known as **"microsatellite mediated enhancement"** (**MME**) of gene transcription (Kumar et al. 2016).

References

Litt M, Luty JA (1989) A hypervariable microsatellite revealed by in–vitro amplification of a dinucleotide repeat within the cardiac nuscle actin gene. Am J Hum Genet 44:397–401.

Zane L, Bargelloni L, Patarnello T (2002) Stretegy for microsatellite isolation: a review. Mol Ecol 11: 1–16.

Li Y–C, Abraham Korol B, Fahima I, Nevo E (2004) Microsatellites Within Genes: Structure, Function, and Evolution. Mol Bio and Evo 2(6): 991–1007.

Fujimori S, Washio T, Higo K, Ohtomo Y, Murakami K, Matsubara K, Kawai J, Carninci P, Hayashizaki Y, Kikuchi S, Tomita M (2003) A novel feature of microsatellites in plants: a distribution gradient along the direction of transcription. FEBS Lett 554: 17–22.

Simpson P, Ayyar S (2008) Evolution of *cis*–regulatory sequences in *Drosophila*. In: van Heyningen V, Hill RE (eds) Advances in genetics (Long range control of gene expression), vol 61. Elsevier, pp 67–106.

Biemont C, Vieira C (2006) Genetics: junk DNA as an evolutionary force. Nature 443: 521–524.

Morgante M, Hanafey M, Powell W (2002) Microsatellites are preferentially associated with nonrepetitive DNA in plant genomes. Nat Genet 30: 194–200.

Lawson MJ, Zhang L (2006) Distinct pattern of SSR distribution in the *Arabidopsis thaliana* and rice genomes. Genome Biol 7: R14.

Burke TW, Kadonaga JT (1997) The downstream core promoter element, DPE, is conserved from *Drosophila* to humans and is recognized by TAFII60 of Drosophila. Genes Dev 11: 3020–3031.

Anish R, Hossain MB, Jacobson RH, Takada S (2009) Characterization of transcription from TATA–less promoters: identification of a new core promoter element XCPE2 and analysis of factor requirements. PLoS One 4: e5103.

Lim CY, Santoso B, Boulay T, Dong E, Ohler U et al (2004) The MTE, a new core promoter element for transcription by RNA polymerase II. Genes Dev 18: 1606–1617.

Kashi Y, King D, Soller M (1997) Simple sequence repeats as a source of quantitative genetic variation. Trends Genet. 13:74–78.

Trifonov E. N. (2003) Tuning function of tandemly repeating sequences: a molecular device for fast adaptation. Pp. 1–24 in S. P. Wasser, ed., Evolutionary theory and processes: nodern horizons, papers in honor of Eviatar Nevo. Kluwer Academic Publishers. Amsterdam, The Netherlands.

Jackson AL, Chen R, Loeb LA (1998) Induction of microsatellite instability by oxidative DNA damage. Proc Natl Acad Sci USA 95: 12468–12473.

Young ET, Sloan JS, van Riper K (2000) Trinucleotide repeats are clustered in regulatory genes in *Saccharomyces cerevisiae*. Genetics 154: 1053–1068.

Van der Heijden R, Jacobs DI, Snoeijer W, Hallard D, Verpoorte V (2004) The *Catharanthus* alkaloids: pharmacognosy and biotechnology. Curr Med Chem 11: 607–628.

Lawson MJ, Zhang L (2006) Distinct pattern of SSR distribution in the *Arabidopsis thaliana* and rice genomes. Genome Biol 7: R14.

Contente A, Dittmer A, Koch MC, Roth J, Dobbelstein M (2002) A polymorphic microsatellite that mediate induction of *PIG3* by p53. Nat Genet 30: 315–320.

Albanese V, Biguet, NF, Kiefer H, Bayard E, Mallet J, Meloni R (2001) Quantitative effects on gene silencing by allelic variation at a tetranucleotide microsatellite. Hum. Mol. Genet. 10: 1785–1792.

Toth G, Gaspari Z, Jurka J (2000) Microsatellites in different eukaryotic genomes: survey and analysis. Genome Res 10: 967–981.

Sangwan I, O'Brian MR (2002) Identification of a soybean protein that interacts with GAGA element dinucleotide repeat DNA. Plant Physiol 129: 1788–1794.

Pauli S, Rothnie HM, Chen G, He X, Hohn T (2004) The cauliflower mosaic virus 35S promoter extends into the transcribed region. J Virol 78: 12120–12128.

Bao S, Corke H, Sun M (2002) Microsatellites in starch–synthesizing genes in relation to starch physicochemical properties in waxy rice (*Oryza sativa* L.). Theor Appl Genet 105: 898–905.

Zain R, Sun J–S (2003) Do natural DNA triple–helical structures occur and function in vivo? CMLS, Cell Mol Life Sci 60: 862–870.

Lu Q, Teare JM, Granok H, Swede MJ, Xu J, Elgin SC (2003) The capacity to form H–DNA cannot substitute for GAGA factor binding to a (CT)n.(GA)n regulatory site. Nucleic Acids Res 31:2483–2494.

Martin P, Makepeace K, Hill SA, Hood DW, Moxon ER (2004) Microsatellite instability regulates transcription factor binding and gene expression. Proc Natl Acad Sci USA, 102: 3800–3804.

Busturia A, Lloyd A, Bejarano F, Zavortink M, Xin H, Sakonju S (2001) The MCP silencer of the *Drosophila* Abd–B gene requires both pleiohomeotic and GAGA factor for the maintenance of repression. Development 128: 2163–2173.

Sangwan I, O'Brian MR (2002) Identification of a soybean protein that interacts with GAGA element dinucleotide repeat DNA. Plant Physiol 129: 1788–1794.

Santi L, Wang Y, Stile MR, Berendzen K, Wanke D, Roig C, Pozzi C, Muller K, Muller J, Rohde W, Salamini F (2003) The GA octodinucleotide repeat binding factor BBR participates in the transcriptional regulation of the homeobox gene Bkn3. Plant J 34: 813–826.

Kooiker M, Airoldi CA, Losa A, Manzotti PS, Finzi L, Kater MM, Colombo L (2005) BASIC PENTACYSTEINE1, a GA binding protein that induces conformational changes in the regulatory region of the homeotic *Arabidopsis* gene SEEDSTICK. Plant Cell 17: 722–729.

Hulzink RJ, de Groot PF, Croes AF, Quaedvlieg W, Twell D, Wullems GJ, Van Herpen MM (2002) The 5'–untranslated region of the ntp303 gene strongly enhances translation during pollen tube growth, but not during pollen maturation. Plant Physiol 129: 342–353.

Wells RD, Collier DA, Hanvey JC, Shimizu M, Wohlrab F (1998) The chemistry and biology of unusual DNA structures adopted by oligopurine′ oligopyrimidine sequences. FASEB J 2: 2939–2949.

Glaser RL, Thomas GH, Siegfried ES, Elgin SCR, Lis JT (1990) Optimal heat–induced expression of the *Drosophila* hsp26 gene requires a promoter sequence containing (CT)n.(GA)n repeats. J Mol Biol 211: 751–761.

Htun H, Dahlberg JE (1989) Topology and formation of triple stranded–H–DNA. Science 243: 1571–1576.

Sandaltzopoulos R, Mitchelmore C, Bonte E, Wall G, Becker PB (1995) Dual regulation of the Drosophila hsp26 promoter *in vitro*. Nucleic Acids Res 23: 2479–2487.

Sharopova N (2008) Plant simple sequence repeats: distribution, variation, and effects on gene expression. Genome 51: 79–90.

Kumar S, Bhatia S. (2016)A polymorphic (GA/CT)*n*- SSR influences promoter activity of *Tryptophan decarboxylase* gene in *Catharanthus roseus*L. Don. *Scientific Reports* vol. 6 (33280).

Chapter - 7

Legume Breeding for Fungal Resistance: A lesson to Learn

—Baljinder Singh, Sahil Mehta, Manish Tiwari and Sabhyata Bhatia

7.1 Introduction

In nature, crop plants especially legumes are continuously exposed to many environmental stresses (individual and combinatorial stresses) at a time. Among all, *biotic stresses have a major negative impact on crop production worldwide.* In agriculture, crops especially legumes are attacked by many fungi, insects, bacteria, nematodes, weeds, parasites, rodents, birds and viruses. They affect plant growth, development, and health which lead to severe decline in yield, vigour and productivity due to imbalance at cellular, molecular, and physiological levels. This issue needs a paradigm shift in the breeding strategies to strengthen our traditional crop improvement programmes.

Considering the present scenario, it is imperative to switch to sophisticated methods that combat biotic environmental challenges and improve crop yield at a same time. Among these, marker-assisted breeding is a commonly utilized method for enhancing biotic stress tolerance in crops especially legumes (Barilli et al. 2010; Diers et al. 2014; Liu et al. 2016). It is an economical and environment friendly method. Marker assisted breeding coupled with progress in agricultural technology has made remarkable progress in increasing crop yields and feeding tremendous population. At present, the world population is around 7.5 billion and increasing bit-by-bit with each day. It has been reported to reach approximately 10 billion by the year 2056 (United Nations projections). All over the globe, this enormous population consumes mainly cereals, pulses, oilseeds, etc.

Among all, legumes are the largest major food crops consumed worldwide after cereals (Kudapa et al. 2013). Legumes are the members of Fabaceae/Leguminosae family and accounts for approximately 750 genera and 19500 species. On the basis of area harvested and total production, legumes account 27 per cent approximately in global agricultural crop production (Graham and Vance 2003). Annual grain legume production area and total production worldwide is 78 MHa and 70 MTonnes respectively (FAO, 2012). In both developed and developing countries, some legumes are the important component of human diet which includes soybean (*Glycine max*), groundnut (*Arachis hypogaea*), Lupins (*Lupinus spp.*), cowpea (*Vigna unguiculata*), lentil (*Lens culinaris*), faba bean (*Vicia faba*), common bean (*Phaseolus vulgaris*), earth pea (*Vigna subterranea*), pigeonpea (*Cajanus cajan*), common vetch (*Vicia sativa*), pea (*Pisum sativum*), chickpea (*Cicer arietinum*), mungbean (*Vigna radiata*) and lupin (*Lupinus luteus*) (Kudapa et al. 2013).

On the basis of their ability to grow in different seasons, legumes are grouped into two groups – (1) cool season legumes and (2) warm season legumes (Toker and Yadav 2010). The four major cool season legumes are lentil, chickpea, dry pea and broad bean (Latef and Ahmad 2015). Examples of warm season legumes include soybean, pigeon pea, mungbean and cowpea. Apart from human consumption, they are also beneficiary for environment. They are capable of interacting symbiotically with differentiated forms of soil-borne bacteria, the *Rhizobium*, which helps plant to fix atmospheric nitrogen and increase the soil quality. Legumes and cereals are often grown in rotation to each other because of the capability of legumes to fix atmospheric nitrogen.

In developing countries, legumes play an important role in human diet; that's why they are often referred as 'poor man's meat'. They are generally good source of proteins, complex carbohydrates, dietary fibers, oils, vitamin-B, vitamin-C, vitamin-E and many minerals. Different legumes vary from each other in their nutritional value-dietary proteins, vegetable oils, total protein content, specific amino acids content, etc. They provide around 20-40 per cent of all dietary proteins (Zhu et al. 2005) to vegetarians. Within legumes, soybean and peanut are main sources for 35 per cent of world's processed vegetab. Protein content of legumes varies from 18-25 per cent as in case of peas and beans to 35-42 per cent in case of soybean (Table 7.1), which is comparable to protein content of meat (18-25 per cent) but is contrasting to protein content of cereals (7-13 per cent) (Bouchenak and Lamri-Senhadji 2013). Legume proteins are low in sulphur containing amino acids like methionine, tryptophan and cysteine but they have high amount of lysine if compared to cereals. So, a diet with combination of both kinds of seeds would provide better protein quality than either of them (Young and Pellett 1994). Types of B vitamins present in legumes are thiamin, folic acid, niacin, riboflavin and pyridoxine. These vitamins play important role in fatty acid metabolism pathways (Rebello et al. 2013).

Table 7.1: Macronutrient information of legumes.

Sl. No.	*Scientific name*	*Common name*	*Carbohyd-rates (%)*	*Proteins (%)*	*Fat (%)*	*Fiber (%)*	*References*
1	*Cicer arietinum*	Chickpea	63.4	20.7	4.16	2.49	Bravo et al. (1999)
2	*Lens culinaris*	Lentil	56.4	20.6	2.15	6.83	De Almeida Costa et al. (2006)
3	*Pisum sativum*	Garden pea	52.5	21.9	2.34	10.4	De Almeida Costa et al. (2006)
4	*Cajanus cajan*	Pigeon pea	62.7	19.3	2.0	6.4	Singh and Sing (1992)
5	*Phaseolus vulgaris*	Common bean	52.5	20.9	2.49	8.55	De Almeida Costa et al. (2006)
6	*Glycine max*	Soybean	35	36	19	17	Mateos-Aparicio et al. (2008)
7	*Vicia faba*	Faba bean	58	26	1-2	25	US Department of Agriculture, 2001

Table 7.2: Summary of QTLs identified for rust tolerance in some legume crop species.

Sl. No.	*Rust species*	*Plant*	*Population*	*Parental cross*	*QTL/gene(s)*	*References*
1.	*Uromyces pisi* (Pers.) Wint.	*Pisum fulvum* L.	F_2	*P. fulvum* (IFPI3260, Resistant) X *P. fulvum* (IFPI3251, Susceptible)	*QTL (Up1)*	Barilli et al. (2010)
2.	*Uromyces fabae* (Pers.) de-Bary	*Pisum sativum* L.	F_6 & F_7 (RILs)	*P. sativum* (HUVP1, Susceptible) X *P. sativum* (FC1, Resistant)	*Qruf*	Rai et al. (2011)
3.	*Uromyces fabae* (Pers.) de-Bary.	*Pisum sativum* L.	31 BC_1F_1	*P. sativum* (HUVP 1, Susceptible) X *P. sativum* (FC 1, Resistant)	*Ruf*	Vijayalakshmi et al. (2005)
4.	*Uromyces pisi* (Pers.) Wint.	*Pisum fulvum* L.	F_7 (RILs)	*P. fulvum* (IFPI3260, Resistant) X *P. fulvum* (IFPI3251, Susceptible)	*UpDSII*, *UpDSIV and UpDSIV.2*	Barilli et al. (2018)
5.	*Phakopsora pachyrhizi*	*Glycine max L. Merill*	F_2 & F_3	*G. max* (Prichard, Susceptible) X *G. max* (PI 567068A, Resistant)	*Rpp6/Rpp[PI567068A]* haplotype	King et al. (2016)
6.	*Phakopsora pachyrhizi* Syd.	*Glycine max L. Merill*.	F_2 & F_3	*G. max* (DS-880, Susceptible) X *G. max* (PI 567102B, Resistant)	*Rpp6*	Li et al. (2012)
7.	*Phakopsora pachyrhizi* Syd.	*Glycine max L. Merill*	F_5 & F_6 (RILs)	*G. max* L. (IAR 2001 BSR, Susceptible) X *G. max* L. (PI 567104B, Resistant)	*Rpp*	Liu et al. (2016)
8.	*Phakopsora pachyrhizi* Syd.	*Glycine max L. Merill*	Three F_2 populations	*G. max* L. (Zhongdou 40, Susceptible) X *G. max* L. (SX6907, Resistant); *G. max* L. (Tianlong 1, Susceptible) X *G. max* L. (SX6907, Resistant); *G. max* L. (Pudou11, Susceptible) X *G. max* L. (SX6907, Resistant)	*Rpp6907*	Chen et al. (2015)
9.	*Phakopsora pachyrhizi* Syd.	*Glycine max L. Merill*	F_3	*G. max* (LG01-5087-9, Susceptible) and *G. max* (PI 30970, Resistant)	*Rpp2*	Yu et al. (2015)

(Contd...)

10.	*Phakopsora pachyrhizi* Syd.	*Glycine max L. Merill*	F_2	*G. max* (PI120049, Resistant) X *G. max* (Tainung3 and Tainung 4, Susceptible)	*Rpp1*	McLean and Byth (1980)
11.	*Phakopsora pachyrhizi* Syd.	*Glycine max L. Merill*	F_2	*G. max* (PI230970, Resistant) X *G. max* (Centennial, Susceptible); *G. max* (PI230971, Resistant) X *G. max* (D75-10169, Susceptible)	*Rpp2*	Bromfield and Hartwig (1980)
12.	*Phakopsora pachyrhizi* Syd.	*Glycine max L. Merill*	F_2 & F_3	*G. max* (Williams 82, Susceptible) X *G. max* (PI462312, Resistant)	*Rpp3*	Hyten et al. (2009)
13.	*Phakopsora pachyrhizi* Syd.	*Glycine max L. Merill*	F_3	*G. max* (PI459025, Resistant) X *G. max* (PI 200492, PI 230970, and PI462312, Susceptible)	*Rpp4*	Hartwig (1986)
14.	*Phakopsora pachyrhizi* Syd.	*Glycine max L. Merill*	5 F_2 populations	*G. max* (CD 208, Susceptible) X *G. max* (PI 200487, PI 200526, PI230970, PI459025, PI 471904, Resistant)	*Rpp5*	Gracia et al. (2008)
15.	*Uromyces ciceris-arietini*	*Cicer reticulatum*	F_6 & F_7 (RILs)	*Cicer arietinum* (ILC72, Susceptible) X *Cicer reticulatum* (Cr5-10, Resistant)	*Uca1/uca1*	Madrid et al. (2008)
16.	*Uromyces viciafabae* Pers.	*Lens culinaris* Medik.	RILs	*L. culinaris* (ILL-4605, Resistant) X *L. culinaris* (ILL-5888, Susceptible)	*F7XEM4a*	Saha et al. (2010)
17.	*Uromyces appendiculatus* Pers.: Pers. Unger	*Phaseolus vulgaris* L.	F_2	*P. vulgaris* (Kranskop, Resistant) X *P. vulgaris* (cultivar Bonus, Susceptible)	*Ur-13*	Mienie et al. (2005)
18.	*Uromyces appendiculatus* Pers.: Pers. Unger	*Phaseolus vulgaris* L.	F_2	*P. vulgaris* (GN1140, Resistant) X *P. vulgaris* L. (GN Nebr. #1, Susceptible)	*Ur-7*	Park et al. (2003)
19.	*Uromyces appendiculatus* Pers.: Pers. Unger	*Phaseolus vulgaris* L.	F_2	*P. vulgaris* (Olathe, Resistant) X *P. vulgaris* (Nebr.#1 sel.27, Susceptible)	*Ur-6*	Park et al. (2004)

(Contd...)

20.	*Uromyces appendiculatus* Pers.: Pers. Unger	*Phaseolus vulgaris* L.	F_2	*P. vulgaris* (PC-50, Resistant) X *P. vulgaris* (Chichara 83-109, Susceptible)	*Ur-9*	Park et al. (1999)
21.	*Uromyces appendiculatus* Pers.: Pers. Unger	*Phaseolus vulgaris* L.	68 RILs	*P. vulgaris* (PC-50, Resistant) X *P. vulgaris* (XAN-159, Susceptible)	*Ur-12*	Jung et al. (1999)
22.	*Uromyces appendiculatus* Pers.: Pers. Unger	*Phaseolus vulgaris* L.	F_2 & F_3	*P. vulgaris* (Ouro Negro, US Pinto 111, Susceptible) X *P. vulgaris* (AxS 37, CNC, CSW 643, Resistant)	*Ur-14*	De Sauza et al. (2011)
23.	*Uromyces viciae-fabae* (Pers.) J. Short.	*Vicia faba*	F_2	*V. faba* (2N52, Resistant) X *V. faba* (VF-176, Susceptible)	*Uvf-1*	Avila et al. (2003)
24.	*Puccinia arachidis* Speg.	*Arachis hypogaea* L.	268 F_7 RILs	*A. hypogaea* (TAG 24, Susceptible) X *A. hypogaea* (GPBD 4, Resistant)	*QTLrust01*	Khedikar et al. (2010)

Table 7.3: Summary of QTLs identified for ascochyta blight tolerance in some legume crop species.

Sl. No.	Pathogen	Plant	Population	Parental cross	QTL/gene	References
1	*Didymella pinodes*	*Pisum sativum L.*	F_2	*P. sativum* (3148-A88, 3146-A26, Resistant) X *P. sativum* (Rovar, Susceptible)	*Asc1.1, Asc2.1, Asc3.1, Asc4.1, Asc7.2*	Timmerman-Vaughan et al. (2016)
2	*Mycosphaerella pinodes*	*Pisum sativum* L.	F_2 & F_6 (RILs)	*P. sativum* (DP, partially Resistant) X *P. sativum* (JI296, Susceptible)	*QTL mpIII-1, QTL mpIII-4, QTL mpVII-1*	Prioul-Gervais et al. (2007)
3	*Didymella pinodes*	*Pisum sativum* L.	111 F_6 & F_7 (RILs)	*P. sativum* ssp. *Syriacum* (P665, Resistant) X *P. sativum ssp. sativum cv.* (Messire, Susceptible)	*MpIII.2, MpIII.4, MpII.1_DRseedl, MpVI.1, MpII.1, MpIII.5, MpV.3; MpV.2*	Carrillo et al. (2014)
4	*Mycosphaerella pinodes*	*Pisum sativum* subsp. *syriacum*	111 F_6 & F_7 (RILs)	*P. sativum* ssp. *Syriacum* accession (P665) (Resistant) X *P. sativum* ssp. *sativum* cv. (Messire, Susceptible)	*MpIII.1, MpII.1, MpIII.2 and MpIII.3, MpIV.1, MpV.1*	Fondevilla et al. (2008)
5	*Peyronellaea pinodes*	*Pisum* sp.	144 RILs	*P. fulvum* (P651, Resistant) X *Pisum sativum* (Alfetta, Susceptible)	*abIII-1 and abI-IV-2*	Jha et al. (2016)
6	*Peyronellaea pinodes*	*Pisum* sp.	RILs	*P. fulvum* (P651, Resistant) X *Pisum sativum* (Alfetta, Susceptible)	*abI-IV-2.1 and abI-IV-2.2*	Jha et al. (2017)
7	*Ascochyta rabiei* (Pass.) Labrousse	*Cicer arietinum* L.	104 F_6 & F_7 (RILs)	*C. arietinum* (ILC72, Resistant) X *C. reticulatum* (Cr5-10, Susceptible)	*Ein3, Avr9/Cf9 and Argonaute 4*	Madrid et al. (2014)
8	*Ascochyta rabiei* (Pass.) Labrousse	*Cicer arietinum* L.	F_5 & F_6 (RILs)	*C. arietinum* (FLIP84-92C, Resistant) X *C. reticulatum* (PI 599072, Susceptible)	*QTL-1 and QTL-2*	Santra et al. (2000)
9	*Ascochyta rabiei* (Pass.) Labrousse	*Cicer arietinum* L.	F_7 (RILs)	*C. arietinum* (PI 359075, Susceptible) X *C. arietinum* (FLIP84-92C, Resistant)	*Ar19*	Cho et al. (2004)

(Contd...)

10	*Ascochyta rabiei* (Pass.) Labrousse	*Cicer arietinum* L.	F_6 & F_7 (RILs)	*C. arietinum* (ILC72, Resistant) X *C. reticulatum* (Cr5-10, Susceptible)	*QTL(?)*	Cobos et al. (2006)
11	*Ascochyta rabiei* (Pass.) Labrousse	*Cicer arietinum* L.	F_2	*C. arietinum* (ICC 4991 Pb 7, Susceptible) X *C. arietinum* (ICCV 04516, Resistant)	*QTL1, QTL2* and *QTL3*	Kottapalli et al. (2009)
12	*Ascochyta rabiei* (Pass.) Labrousse	*Cicer arietinum* L.	92 RILs	*C. arietinum* (ICCV 96029, Susceptible) X *C. arietinum* (CDC Frontier, Resistant)	*qtlAb-1.1, qtlAb-2.1, qtlAb-3.1, qtlAb-4.1, qtlAb-6.1, qtlAb-7.1, qtlAb-8.1, qtlAb-8.2* and *qtlAb-8.3*	Daba et al. (2016)
13	*Ascochyta rabiei*	*Cicer arietinum* L.	188 F_8 RILs	*C. arietinum* (JG 62, Susceptible) X *C. arietinum* (ICCV 05530, Resistant)	*AB-Q-SR-4-1* and *AB-Q-APR-4-1*	Garg et al. (2018)
14	*Ascochyta lentis*	*Lens culinaris* Medik.	F_6 (RILs)	*L. culinaris* (Indianhead, Resistant) X *L. culinaris* (Northfield, previously Resistant); *L. culinaris* (Indianhead) X *L. culinaris* (Digger, moderately Resistant); *L. culinaris* (Northfield) X *L. culinaris* (Digger)	*AB_IH1, AB_IH2.1, AB_NF1* and *AB_IH2.2*	Sudheesh et al. (2016)
15	*Ascochyta lentis*	*Lens culinaris* ssp. *culinaris*	F_2 & F_2	*L. culinaris* (ILL7537, Resistant) X *L. culinaris* (ILL5588 and ILL6002, Susceptible)	*QTL-6, QTL-7, QTL-8, QTL-5, QTL-4, QTL-1, QTL-2, QTL-3*	Taylor et al. (2006)
16	*Ascochyta lentis*	*Lens culinaris* Medik.	F_2	*L. culinaris* (Eston, Susceptible) X *L. culinaris* (Indianhead, Resistant)	*UBC2271290 marker linked to ral2*	Chowdhury et al. (2001)
17	*Ascochyta lentis*	*Lens culinaris* Medik.	F_3	*L. culinaris* (ILL5588, Resistant) X *L. culinaris* (ILL6002, Susceptible)	AbR_1	Ford et al. (1999)
18	*Ascochyta fabae* Speg.	*Vicia faba* L.	F_2	*V. faba* (29H, Resistant) X *V. faba* (VF136, Susceptible).	*Af3, Af4, Af5, Af6, Af7* and *Af8*	Avila et al. (2004)
19	*Ascochyta fabae* Speg.	*Vicia faba* L.	F_2	*V. faba* (Vf6, Resistant) X *V. faba* (Vf136, Susceptible)	*Af1* and *AF2*	Román et al. (2003)
21	*Ascochyta fabae* Speg.	*Vicia faba* L.	F_5 & F_6 (RILs)	*V. faba* (Icarus, Susceptible) X *V. faba* (Ascot, Resistant)	*QTL-1, QTL-2, QTL-3* and *QTL-4*	Kaur et al. (2014)
22	*Ascochyta fabae* Speg.	*Vicia faba* L.	119 F_7 & F_8 (RILs)	*V. faba* (29H, Resistant) X *V. faba* (Vf1 36, Susceptible)	*Af2* and *Af3*	Atienza et al. (2016)

Table 7.4: Summary of QTLs identified for wilt tolerance in some legume crop species.

Sl. No.	*Pathogen*	*Plant*	*Population*	*Parental cross*	*QTL/gene*	*References*
1	*Fusarium oxysporum f.* sp. *ciceris*	*Cicer arietinum* L.	94 F_9 & F_{10} (RILs)	*C. arietinum* (JG62, Susceptible) X *C. arietinum* (WR315, Resistant)	*WILT-1* and *WILT-2*	Patil et al. (2014)
2	*Fusarium oxysporum* f. sp. *ciceris*	*Cicer arietinum* L.	188 F_2 & F_3	*C. arietinum* (C214, Susceptible) X *C. arietinum* (WR315, Resistant)	*FW-Q-APR-6-1* and *FW-Q-APR-6-2*	Sabbavarapu et al. (2013)
3	*Fusarium oxysporum* f. sp. *ciceris*	*Cicer arietinum* L.	F_6 & F_7 (RILs)	*C. arietinum* (CA2156, Susceptible) X *C. arietinum* (JG62, Resistant); *C. arietinum* (CA2139, Resistant) X *C. arietinum* (JG62, Resistant)	*Foc01/foc01*	Cobos et al. (2005)
4	*Fusarium oxysporum* f. sp. *ciceris* race 3	*Cicer arietinum* L.	F_7 (RILs)	*C. arietinum* (WR315, Resistant) X *C. arietinum* (C104, Susceptible)	*foc-3*	Sharma et al. (2004)
5	*Fusarium oxysporum* f. sp. *ciceris* race 4 and 5	*Cicer arietinum* L.	130 F_7 (RILs)	*C. arietinum* (ICC-4958, Resistant) X *C. reticulatum* (PI 498777, Susceptible)	*Foc4* and *Foc5*	Winter et al. (2000)
6	*Fusarium oxysporum* f. sp. *ciceris* race 0	*Cicer arietinum* L.	F_6 & F_7 (RILs)	*C. arietinum* (CA2156, Susceptible) X *C. arietinum* (JG62, Resistant); *C. arietinum* (CA2139, Resistant) X *C. arietinum* (JG62, Resistant)	*Foc02/foc02*	Halila et al. (2009)
7	*Fusarium oxysporum* f. sp. *ciceris*	*Cicer arietinum* L.	188 F_8 (RILs)	*C. arietinum* (JG 62, Susceptible) X *C. arietinum* (ICCV 05530, Resistant)	FW-Q-APR-2-1, FW-Q-APR-2-1 and FW-Q-APR-4-1	Garg et al. (2018)
8	*Fusarium udum*	*Cajanus cajan* L. Millsp.	RILs	*C. cajan* (CPL20096, Resistant) X *C. cajan* (ICPL 332, Susceptible)	*C.cajan_03203*	Singh et al. (2016)

(Contd...)

9	*Fusarium udum*	*Cajanus cajan* L. Millsp.	188 RILs	*C. cajan* (ICPL 20096, Resistant) X *C. cajan* (ICPL 332, Susceptible); *C. cajan* (ICPB 2049, Susceptible)X *C. cajan* (ICPL 99050, Resistant)	*qFW11.1, qFW11.2 and qFW11.3*	Saxena et al. (2017)
10	*Fusarium oxysporum* f.sp. *pisi*	*Pisum sativum* L.	187 F_7 (RILs)	*P. sativum* (Shawnee, Resistant) X *P. sativum* (Bohatyr, Susceptible)	*Fnw 4.1, Fnw3.1* and *Fnw 3.2*	McPhee et al. (2012)
11	*Fusarium oxysporum* Schlecht. f.sp. *pisi* (Fop)	*Pisum sativum* L.	80 F_8 RILs	*P. sativum* (Green Arrow, Resistant) X *P. sativum* (PI 179449, Susceptible)	Fw	Kwon et al. (2013)
12	*Fusarium oxysporum* f.sp. *tracheiphilum*	*Vigna unguiculata* L.	F_{10} (RILs)	*V. unguiculata* (California Blackeye 27, Resistant) X *V. unguiculata* (C93W-24-125B-1, Susceptible)	*Fot3-1*	Pottorff et al. (2012)
13	*Fusarium oxysporum* f.sp. *tracheiphilum*	*Vigna unguiculata* L.	F_{10} (RILs)	*V. unguiculata* (IT93K-503-1, Resistant) X *V. unguiculata* (CB46, Susceptible); *V. unguiculata* (CB27, Resistant) X *V. unguiculata* (24-125B-1, Susceptible)	*Fot4-1* and *Fot4-2*	Pottorff et al. (2014)
14	*Fusarium oxysporum* f.sp. *lentis*	*Lens culinaris* Medik.	F_6 & F_8 and F_6 & F_9 (RILs)	*L. culinaris* (ILL5588, Resistant) X *L. culinaris* [L692-16-1(s), Susceptible]	Fw	Eujayl et al. (1998)

Legumes also produce many metabolites and phytochemicals in order to protect them against pathogen and pest attack. Phytochemicals that are present in legumes include enzyme inhibitors, flavonoids, isoflavonoids, polyphenols, lectins, phytoestrogens and saponins. These phytochemicals are capable of preventing various metabolic diseases like diabetes (Jenkins et al. 1981; Kaushik et al. 2010) and coronary heart disease (Anderson et al. 2000), cancer (Kennedy 1995) in lowering of serum cholesterol (Duane, 1997). Phytochemicals protect body tissues from any oxidative damage, they act as antioxidants. Flavonoids act against free radicals, inflammation and platelet aggregation. Phytic acid is capable of controlling cell cycle of malignant cells to prevent uncontrolled cell division and often it is capable of converting them back to their normal state. A special class of chemicals, saponins binds with cholesterol and form an insoluble complex thereby preventing their absorption in small intestine. Vitamin E is also an antioxidant; it prevents membrane lipids oxidation by free radicals (Bouchenak and Lamri-Senhadji 2013). Besides these benefits of phytochemicals certain enzyme inhibitors are anti-nutritional as they inhibit activity of some digestive enzymes. However, these proteins become inactivated after cooking.

Legumes are very important part of animal diet including humans but productivity of these crops is still low due to various biotic and abiotic stresses. This plethora set of stresses limits the legumes productivity. Among multiple stresses, biotic stresses are one of the most serious production constraint in legume productivity. They are projected to worsen with climate change. Within biotic stress, fungal pathogens are the most devastating biotic stress to the legume crops productivity. There are about 50,000 species of fungi that attack a wide range of legumes worldwide. These fungal species infect both grain and forage legumes. Some of major fungal infections like Fusarium wilt, Ascochyta blight and rust have always been major constraints on legume productivity. Identifying QTLs responsible for resistance in legumes for these fungal pathogens and using molecular breeding programs to develop improved tolerant varieties can be a major step to prevent yield loss in legumes. Recent advances in Next Generation Sequencing technologies have accelerated large scale generation of genomic resources and isolation of many QTLs. This book-chapter summarizes QTLs identified in legumes for major fungal stress-rust, Ascochyta blight and Fusarium wilt and focuses on the few developments made on this aspect over the years additionally.

7.2 Biotic stress in legumes

7.2.1 Rust

One of the major fungal stresses in legumes is rust. Both disease and its pathogen are called rust. Approximately 5000 species of rust attack various crop plants globally (Sillero et al. 2006). Mostly rust pathogens are from genus *Uromyces* but some are also from *Puccinia* and *Phakopsora*. Mostly rust infection in legumes start near the end of the season which has low effect on yield loss but if infection starts early in the season then yield loss is very high.

Fungicide application is a good method to stop infection but due to its effect on environment and high cost this method becomes unfavorable in developing countries. The most cost effective method for the control of rust is breeding for rust resistance trait. In economically important legumes like soybean and common bean, resistance reactions to rust have been described as completely monogenic. These crops have been studied extensively for the identification of rust resistance genes and their closely associated markers. The markers identified from these studies are readily being used in marker assisted breeding (Hyten et al. 2009; de Souza et al. 2011). Initially, McLean and Byth (1980) conducted a study to determine the inheritance of rust resistance in three different accessions of soybean and identified single dominant gene (*Rpp1*) for resistance to rust. Till now 6 different resistant loci (*Rpp*) for *P. pachyrhizi* have been identified in soybean, named as: *Rpp1, Rpp2, Rpp3, Rpp4, Rpp5* and *Rpp6* (McLean and Byth 1980; Bromfield and Hartwig 1980; Hyten et al. 2009; Hartwig 1986; Garcia et al. 2008). Introgressions of *Rpp* genes through breeding into elite germplasm have been successful. Diers et al. (2014) have been able to develop eight backcross lines by introducing four different *Rpp* loci into elite lines. Various rust resistance genes have been identified in common bean namely *Ur-3*, *Ur-4*, *Ur-5*, *Ur-6* (Park et al. 2004), *Ur-7* (Park et al. 2003), *Ur-9* (Park et al. 1999), *Ur-11*, *Ur-12* (Jung et al. 1998), *Ur-*

13 (Mienie et al. 2005), Ur-14 (de Souza et al. 2011). While in cool season legume crops rust resistance reactions are incomplete (Rubiales et al. 2011). Various QTL mapping studies have been performed in cool season legumes for rust resistance but in-depth information for resistance genes and their associated markers is not available in most of the cases. List of all the QTLs and genes identified in legumes for rust resistance are given in table 7.2.

7.2.2 *Aschochyta* blight

Ascochyta is a pathogenic fungus to various plants mostly legumes and cereals. In legumes they mostly affect cool season legumes. The most important foliar diseases worldwide are ascochyta blights. Different pathogens are responsible for ascochyta blight in various legumes such as *A. lentis* in lentil, *A. fabae* in faba bean, *Ascochyta rabiei* in chickpea and *Mycosphaerella pinodes* and *A. pisi*, *Phoma medicaginis* var. *pinodella* in pea (Tivoli *et al.*2006). Symptoms of this disease include development of lesions on the surface of leaves, pods and stems. Prolong infection can lead to plant death and reduction in yield. Various methods have been developed for the control of ascochyta blight such as chemical control and cultural practices. However, very less success have been achieved through these methods as most of these strategies are uneconomical and hard to achieve. The most economical method for the control of ascochyta is through marker assisted breeding for resistance.

Large number of QTL mapping studies have been done to identify regions of genome that are responsible for ascochyta blight resistance in chickpea (Madrid *et al.,* 2014; Santra *et al.,* 2000; Cobos *et al.,* 2006; Daba *et al.,* 2016; Garg *et al.,* 2018), lentil (Sudheesh *et al.,* 2016; Rubeena *et al.,* 2006; Chowdhary *et al.,* 2001; Ford *et al.,* 1999), faba bean (Román et al., 2003; Avila et al., 2004; Rashid *et al., 1991;* Kaur et al., 2014; Atienza et al., 2016) and field pea (Timmerman-Vaughan et al., 2016; Prioul et al., 2007; Carrillo *et al.,* 2014; Fondevilla et al.,2008; Jha et al., 2016, 2017). However, genomic region covered by QTLs identified in pea is large which is not suitable for marker-assisted selection. Most of the QTLs/gene(s) that are identified till now are mentioned in table 7.3.

7.2.3 *Fusarium* wilt

Fusarium wilt is a vascular wilt disease caused by soil borne fungal pathogen *Fusarium oxysporum*. It is one of the major biotic stress that reduces yield of forage and grain legumes around the world. The species of *Fusarium oxysporum* is further divided into *forma specialis* on the basis of the host plants they infect. On the basis of genetic and molecular characterization different pathotypes and races have been identified for most of the *forma specialis*. Till date, eight races and two pathotypes of *F. oxysporum* f. sp. *ciceris* have been identified for chickpea (*Cicer arietinum* L.) (Jiménez-Fernández et al. 2013), seven races have been described of *F. oxysporum* f. sp. *phaseoli* for common bean (*Phaseolus vulgaris* L.*)* (Alves-Santos et al. 2002), four races of *Fusarium oxysporum* f.sp. *tracheiphilum* have been identified for cowpea (*Vigna unguiculata* (L) Walp.) (Pottorff et al. 2012) and four races of *F. oxysporum* f. sp. *Pisi have been identified for* pea (*Pisum sativum* L.) (Rispail and Rubiales 2013).

The pathogen enter through the root into the vascular tissue of the plant and is responsible for chlorosis and wilting of older leaves. Continuous progression of infection leads to wilting of younger leaves also and eventually stunting of entire plant. As in case on many pathogenic soil borne fungi, disease control is difficult using chemical methods. So, developing resistant varieties through breeding is a good method for disease control. Qualitative monogenic resistance have been already described for fusarium wilt in most of the legumes (Infantino et al. 2006) but pathogen can easily overcome this resistance by emergence of new strains. Using quantitative traits for breeding purposes will improve the durability of resistance.

Batista et al. (2017) identified that resistance to fusarium wilt in common bean is a dominant trait which is controlled by few major genes and polygenes. In *Vigna unguiculata* two independent resistance loci *Fot4-1* and *Fot4-2* and one locus *Fot 3-1* have been identified for race 4 and race 3 respectively (Pottorff et al. 2014; Pottorff et al. 2012). In case of chickpea, resistance to *Fusarium oxysporum f.* sp. *Ciceris* is either monogenic or polygenic depending upon the race. List of all the QTLs and genes identified for wilt resistance in legumes is given in table 7.4.

7.3 Conclusions and future prospects

Being a short duration crop, legumes grow in a wide range of environment. Adding to this advantage, it is a bonus to farmers of agricultural lands which remain fallow for two to three months after the harvest of the main crops. However, being sensitive to both abiotic and biotic stresses, its widespread adoption is not gaining ground. Therefore, there is an urgent need to evaluate the wild relatives of many legumes germplasm and identify genes for resistance to both stresses. Incorporation of genes for resistance against biotic stress from closely related species should be the top priority for current breeders. In this regard, major focus should be on field screening as it has a direct use in the improvement of legumes to biotic stresses.

Many marker linked genes for rust, blight and wilt resistance have already been identified. This directly highlights the value of marker assisted breeding technique in the improvement of legume tolerance to biotic stresses. In future, more robust markers will be identified in legumes like common bean, mungbean, black gram, lentil, cowpea, pigeon pea, azuki bean, chicken pea, etc. This will lead to cross species transfer of markers from model to various pulses in which genomic resources development is slow.

References

Alves-Santos FM, Cordeiro-Rodrigues L, Sayagués JM, Martín-Domínguez R, García-Benavides P, Crespo MC et al. (2002) Pathogenicity and race characterization of *Fusarium oxysporum* f. sp. phaseoli isolates from Spain and Greece. Plant Pathology 51:605-611.

Anderson JW, Hanna TJ, Peng X, Kryscio RJ (2000) Whole grain foods and heart disease risk. Journal of the American College of Nutrition 19:291S-299S.

Atienza SG, Palomino C, Gutiérrez N, Alfaro CM, Rubiales D, Torres AM et al. (2016) QTLs for ascochyta blight resistance in faba bean (*Vicia faba* L.): validation in field and controlled conditions. Crop and Pasture Science 67:216-224.

Avila CM, Satovic Z, Sillero JC, Rubiales D, Moreno MT, Torres AM (2004) Isolate and organ-specific QTLs for ascochyta blight resistance in faba bean (*Vicia faba* L). Theoretical and Applied Genetics 108:1071-1078.

Avila CM, Sillero JC, Rubiales D, Moreno MT, Torres AM (2003) Identification of RAPD markers linked to the Uvf-1 gene conferring hypersensitive resistance against rust (*Uromyces viciae*-fabae) in *Vicia faba* L. Theoretical and Applied Genetics 107:353-358.

Barilli E, Cobos MJ, Carrillo E, Kilian A, Carling J, Rubiales D (2018) A high-density integrated DArTseq SNP-based genetic map of *Pisum fulvum* and identification of QTLs controlling rust resistance. Frontiers in Plant Science 9:167.

Barilli E, Satovic Z, Rubiales D, Torres AM (2010) Mapping of quantitative trait loci controlling partial resistance against rust incited by *Uromyces pisi* (Pers.) Wint. in a *Pisum fulvum* L. intraspecific cross. Euphytica 175: 151-159.

Batista RO, Silva LC, Moura LM, Souza MH, Carneiro PCS, Filho JLSC et al. (2017) Inheritance of resistance to fusarium wilt in common bean. Euphytica 213:133.

Bouchenak M, Lamri-Senhadji M (2013) Nutritional quality of legumes, and their role in cardiometabolic risk prevention: A review. Journal of Medicinal Food 16:185-198.

Bravo L, Siddhuraju P, Saura-Calixto F (1999) Composition of underexploited Indian pulses. Comparison with common legumes. Food Chemistry 64: 185-192.

Bromfield KR, Hartwig EE (1980) Resistance to soybean rust and mode of inheritance. Crop Science 20:254-255.

Carrillo E, Satovic Z, Aubert G, Boucherot K, Rubiales D, Fondevilla S (2014) Identification of quantitative trait loci and candidate genes for specific cellular resistance responses against *Didymella pinodes* in pea. Plant Cell Reports 33:1133-1145.

Chen H, Zhao S, Yang Z, Sha A, Wan Q, Zhang C et al. (2015) Genetic analysis and molecular mapping of resistance gene to *Phakopsora pachyrhizi* in soybean germplasm SX6907. Theoretical and Applied Genetics 128: 733-743.

Cho S, Chen W, Muehlbauer FJ (2004) Pathotype-specific genetic factors in chickpea (*Cicer arietinum* L.) for quantitative resistance to ascochyta blight. Theoretical and Applied Genetics 109:733-739.

Chowdhury MA, Andrahennadi CP, Slinkard AE, Vandenberg A (2001) RAPD and SCAR markers for resistance to acochyta blight in lentil. Euphytica 118:331-337.

Cobos MJ, Fernández MJ, Rubio J, Kharrat M, Moreno MT, Gil J et al. (2005) A linkage map of chickpea (*Cicer arietinum* L.) based on populations from Kabuli × Desi crosses: location of genes for resistance to fusarium wilt race 0. Theoretical and Applied Genetics 110:1347-1353.

Cobos MJ, Rubio J, Strange RN, Moreno MT, Gil J, Millan T (2006) A new QTL for Ascochyta blight resistance in an RIL population derived from an interspecific cross in chickpea. Euphytica 149:105-111.

Daba K, Deokar A, Banniza S, Warkentin TD, Tar'an B (2016) QTL mapping of early flowering and resistance to ascochyta blight in chickpea. Genome 59:413-425.

de Almeida Costa GE, da Silva Queiroz-Monici K, Pissini Machado Reis SM, de Oliveira AC (2006) Chemical composition, dietary fibre and resistant starch contents of raw and cooked pea, common bean, chickpea and lentil legumes. Food Chemistry 94:327-330.

de Souza TLPO, Dessaune SN, Sanglard DA, Moreira MA, de Barros EG (2011) Characterization of the rust resistance gene present in the common bean cultivar Ouro Negro, the main rust resistance source used in Brazil. Plant Pathology 60:839-845.

Diers BW, Kim K-S, Frederick RD, Hartman GL, Unfried J, Schultz S et al. (2014) Registration of eight soybean germplasm lines resistant to soybean rust. Journal of Plant Registrations 8:96-101.

Eujayl I, Erskine W, Bayaa B, Baum M, Pehu E (1998) Fusarium vascular wilt in lentil: Inheritance and identification of DNA markers for resistance. Plant Breeding 117:497-499.

Fondevilla S, Satovic Z, Rubiales D, Moreno MT, Torres AM (2008) Mapping of quantitative trait loci for resistance to *Mycosphaerella pinodes* in *Pisum sativum* subsp. syriacum. Molecular Breeding 21:439-454.

Ford R, Pang ECK, Taylor PWJ (1999) Genetics of resistance to ascochyta blight (*Ascochyta lentis*) of lentil and the identification of closely linked RAPD markers. Theoretical and Applied Genetics 98:93-98.

Garcia A, Calvo ÉS, de Souza Kiihl RA, Harada A, Hiromoto DM, Vieira LGE (2008) Molecular mapping of soybean rust (*Phakopsora pachyrhizi*) resistance genes: discovery of a novel locus and alleles. Theoretical and Applied Genetics 117:545.

Garg T, Mallikarjuna BP, Thudi M, Samineni S, Singh S, Sandhu JS et al. (2018) Identification of QTLs for resistance to Fusarium wilt and Ascochyta blight in a recombinant inbred population of chickpea (*Cicer arietinum* L.). Euphytica 214:45.

Graham PH, Vance CP (2003) Legumes: importance and constraints to greater use. Plant Physiol. 131:

Halila I, Cobos MJ, Rubio J, Millán T, Kharrat M, Marrakchi M et al. (2009) Tagging and mapping a second resistance gene for Fusarium wilt race 0 in chickpea. European Journal of Plant Pathology 124:87-92.

Hartwig EE (1986) Identification of a fourth major gene conferring resistance to soybean rust1. Crop Science 26:1135-1136.

Hyten DL, Smith JR, Frederick RD, Tucker ML, Song Q, Cregan PB (2009) Bulked segregant analysis using the GoldenGate assay to locate the Rpp3 locus that confers resistance to soybean rust in soybean. Crop Science 49:265-271.

Infantino A, Kharrat M, Riccioni L, Coyne CJ, McPhee KE, Grünwald NJ (2006) Screening techniques and sources of resistance to root diseases in cool season food legumes. Euphytica 147:201-221.

Jenkins DJ, Wolever TM, Taylor RH, Barker H, Fielden H, Baldwin JM et al. (1981) Glycemic index of foods: a physiological basis for carbohydrate exchange. The American Journal of Clinical Nutrition 34:362-366.

Jha AB, Gali KK, Tar'an B, Warkentin TD (2017) Fine mapping of QTLs for Ascochyta blight resistance in pea using heterogeneous inbred families. Frontiers in Plant Science 8:765.

Jha AB, Tar'an B, Stonehouse R, Warkentin TD (2016) Identification of QTLs associated with improved resistance to Ascochyta Blight in an interspecific pea recombinant inbred line population. Crop Science 56:2926-2939.

Jiménez-Fernández D, Landa BB, Kang S, Jiménez-Díaz RM, Navas-Cortés JA (2013) Quantitative and microscopic assessment of compatible and incompatible interactions between chickpea cultivars and *Fusarium oxysporum* f. sp. ciceris races. PLOS ONE 8:e61360.

Jung G, Coyne DP, Bokosi J, Steadman JR, Nienhuis J (1998) Mapping genes for specific and adult plant resistance to rust and abaxial leaf pubescence and their genetic relationships using randomly amplified polymorphic DNA (rapd) markers in common bean. Journal of the American Society for Horticultural Science 123:859-863.

Kaur S, Kimber RBE, Cogan NOI, Materne M, Forster JW, Paull JG (2014) SNP discovery and high-density genetic mapping in faba bean (*Vicia faba* L.) permits identification of QTLs for ascochyta blight resistance. Plant Science 217-218:47-55.

Kaushik G, Satya S, Khandelwal RK, Naik SN (2010) Commonly consumed Indian plant food materials in the management of diabetes mellitus. Diabetes & Metabolic Syndrome: Clinical Research & Reviews 4:21-40.

Kennedy AR (1995) The evidence for soybean products as cancer preventive agents. The Journal of Nutrition 125:733S-743S.

Khedikar YP, Gowda MVC, Sarvamangala C, Patgar KV, Upadhyaya HD, Varshney RK (2010) A QTL study on late leaf spot and rust revealed one major QTL for molecular breeding for rust resistance in groundnut (*Arachis hypogaea* L.). Theoretical and Applied Genetics 121:971-984.

King ZR, Harris DK, Pedley KF, Song Q, Wang D, Wen Z et al. (2016) A novel *Phakopsora pachyrhizi* resistance allele (Rpp) contributed by PI 567068A. Theoretical and Applied Genetics 129:517-534.

Kottapalli P, Gaur PM, Katiyar SK, Crouch JH, Buhariwalla HK, Pande S et al. (2009) Mapping and validation of QTLs for resistance to an Indian isolate of Ascochyta blight pathogen in chickpea. Euphytica 165:79-88.

Kudapa H, Ramalingam A, Nayakoti S, Chen X, Zhuang W-J, Liang X et al. (2013) Functional genomics to study stress responses in crop legumes: progress and prospects. Functional Plant Biology 40:1221-1233.

Kwon SJ, Smýkal P, Hu J, Wang M, Kim S-J, McGee RJ et al. (2013) User-friendly markers linked to Fusarium wilt race 1 resistance Fw gene for marker-assisted selection in pea. Plant Breeding 132:642-648.

Latef AAHA, Ahmad P. 2015. Legumes and breeding under abiotic stress. Legumes under Environmental Stress. John Wiley & Sons, Ltd. p. 1-20.

Li S, Smith JR, Ray JD, Frederick RD (2012) Identification of a new soybean rust resistance gene in PI 567102B. Theoretical and Applied Genetics 125:133-142.

Liu M, Li S, Swaminathan S, Sahu BB, Leandro LF, Cardinal AJ et al. (2016) Identification of a soybean rust resistance gene in PI 567104B. Theoretical and Applied Genetics 129:863-877.

Madrid E, Rubiales D, Moral A, Moreno MT, Millán T, Gil J et al. (2008) Mechanism and molecular markers associated with rust resistance in a chickpea interspecific cross (*Cicer arietinum* × *Cicer reticulatum*). European Journal of Plant Pathology 121:43-53.

Madrid E, Seoane P, Claros MG, Barro F, Rubio J, Gil J et al. (2014) Genetic and physical mapping of the QTLAR3 controlling blight resistance in chickpea (*Cicer arietinum* L). Euphytica 198:69-78.

Mateos-Aparicio I, Redondo Cuenca A, Villanueva-Suárez MJ, Zapata-Revilla MA (2008) Soybean, a promising health source. Nutrición Hospitalaria 23:305-312.

McLean R, Byth D (1980) Inheritance of resistance to rust (*Phakopsora pachyrhizi*) in soybeans. Australian Journal of Agricultural Research 31: 951-956.

McPhee KE, Inglis DA, Gundersen B, Coyne CJ (2012) Mapping QTL for Fusarium wilt Race 2 partial resistance in pea (*Pisum sativum*). Plant Breeding 131:300-306.

Mienie CMS, Liebenberg MM, Pretorius ZA, Miklas PN (2005) SCAR markers linked to the common bean rust resistance gene Ur-13. Theoretical and applied genetics 111:972-979.

Park SO, Coyne DP, Bokosi JM, Steadman JR (1999) Molecular markers linked to genes for specific rust resistance and indeterminate growth habit in common bean. Euphytica 105:133-141.

Park SO, Coyne DP, Steadman JR, Crosby KM, Brick MA (2004) RAPD and SCAR markers linked to the Ur-6 andean gene controlling specific rust resistance in common bean Crop Science 44:1799-1807.

Park SO, Coyne DP, Steadman JR, Skroch PW (2003) Mapping of the Ur-7 gene for specific resistance to rust in common bean. Crop Science 43: 1470-1476.

Patil BS, Ravikumar RL, Bhat JS, Soregaon CD (2014) Molecular mapping of QTLs for resistance to early and late fusarium wilt in chickpea. Czech Journal of Genetics and Plant Breeding 50:171-176.

Pottorff M, Wanamaker S, Ma YQ, Ehlers JD, Roberts PA, Close TJ (2012) Genetic and physical mapping of candidate genes for resistance to *Fusarium oxysporum* f.sp. tracheiphilum race 3 in cowpea [*Vigna unguiculata* (L.) Walp]. PLOS ONE 7:e41600.

Pottorff MO, Li G, Ehlers JD, Close TJ, Roberts PA (2014) Genetic mapping, synteny, and physical location of two loci for *Fusarium oxysporum* f. sp. tracheiphilum race 4 resistance in cowpea [*Vigna unguiculata* (L.) Walp]. Molecular Breeding 33:779-791.

Prioul-Gervais S, Deniot G, Receveur EM, Frankewitz A, Fourmann M, Rameau C et al. (2007) Candidate genes for quantitative resistance to *Mycosphaerella pinodes* in pea (*Pisum sativum* L.). Theoretical and Applied Genetics 114:971-984.

Rai R, Singh AK, Singh BD, Joshi AK, Chand R, Srivastava CP (2011) Molecular mapping for resistance to pea rust caused by *Uromyces fabae* (Pers.) de-Bary. Theoretical and Applied Genetics 123:803-813.

Rebello CJ, Liu AG, Greenway FL, Dhurandhar NV. 2013. Dietary Strategies to Increase Satiety. In: Henry, J., editor). Advances in Food and Nutrition Research. Academic Press. p. 105-182.

Rispail N, Rubiales D (2013) Identification of sources of quantitative resistance to *Fusarium oxysporum* f. sp. medicaginis in *Medicago truncatula*. Plant Disease 98:667-673.

Román B, Satovic Z, Avila CM, Rubiales D, Moreno MT, Torres AM (2003) Locating genes associated with *Ascochyta fabae* resistance in Vicia faba. Australian Journal of Agricultural Research 54:85-90.

Rubiales D, Castillejo MA, Madrid E, Barilli E, Rispail N (2011) Legume breeding for rust resistance: lessons to learn from the model *Medicago truncatula*. Euphytica 180:89-98.

Sabbavarapu MM, Sharma M, Chamarthi SK, Swapna N, Rathore A, Thudi M et al. (2013) Molecular mapping of QTLs for resistance to *Fusarium* wilt (race 1) and Ascochyta blight in chickpea (*Cicer arietinum* L.). Euphytica 193:121-133.

Saha GC, Sarker A, Chen W, Vandemark GJ, Muehlbauer FJ (2010) Identification of markers associated with genes for rust resistance in *Lens culinaris* Medik. Euphytica 175:261-265.

Santra DK, Tekeoglu M, Ratnaparkhe M, Kaiser WJ, Muehlbauer FJ (2000) Identification and mapping of QTLs conferring resistance to ascochyta blight in chickpea. Crop Science 40:1606-1612.

Saxena RK, Singh VK, Kale SM, Tathineni R, Parupalli S, Kumar V et al. (2017) Construction of genotyping-by-sequencing based high-density genetic maps and QTL mapping for fusarium wilt resistance in pigeonpea. Scientific Reports 7:1911.

Sharma KD, Winter P, Kahl G, Muehlbauer FJ (2004) Molecular mapping of *Fusarium oxysporum* f. sp. ciceris race 3 resistance gene in chickpea. Theoretical and Applied Genetics 108:1243-1248.

Sillero JC, Fondevilla S, Davidson J, Patto MCV, Warkentin TD, Thomas J et al. (2006) Screening techniques and sources of resistance to rusts and mildews in grain legumes. Euphytica 147:255-272.

Singh U, Singh B (1992) Tropical grain legumes as important human foods. Economic Botany 46:310-321.

Singh VK, Khan AW, Saxena RK, Kumar V, Kale SM, Sinha P et al. (2016) Nextgeneration sequencing for identification of candidate genes for Fusarium wilt and sterility mosaic disease in pigeonpea (*Cajanus cajan*). Plant Biotechnology Journal 14:1183-1194.

Sudheesh S, Rodda MS, Davidson J, Javid M, Stephens A, Slater AT et al. (2016) SNP-based linkage mapping for validation of QTLs for resistance to ascochyta blight in lentil. Frontiers in Plant Science 7:1604.

Taylor PWJ, Ades PK, Ford R (2006) QTL mapping of resistance in lentil (*Lens culinaris* ssp. culinaris) to ascochyta blight (*Ascochyta lentis*). Plant Breeding 125:506-512.

Timmerman-Vaughan GM, Moya L, Frew TJ, Murray SR, Crowhurst R (2016) Ascochyta blight disease of pea (*Pisum sativum* L.): defence-related candidate genes associated with QTL regions and identification of epistatic QTL. Theoretical and Applied Geñetics 129:879-896.

Toker C, Yadav SS. 2010. Legumes cultivars for stress environments. In: Yadav, S. S. and Redden, R., editors. Climate Change and Management of Cool Season Grain Legume Crops. Dordrecht: Springer Netherlands. p. 351-376.

Vijayalakshmi S, Yadav K, Kushwaha C, Sarode SB, Srivastava CP, Chand R et al. (2005) Identification of RAPD markers linked to the rust (*Uromyces fabae*) resistance gene in pea (*Pisum sativum*). Euphytica 144:265-274.

Winter P, Benko-Iseppon AM, Hüttel B, Ratnaparkhe M, Tullu A, Sonnante G et al. (2000) A linkage map of the chickpea (*Cicer arietinum* L.) genome based on recombinant inbred lines from a *C. arietinum* × *C. reticulatum cross*: localization of resistance genes for fusarium wilt races 4 and 5. Theoretical and Applied Genetics 101:1155-1163.

Young VR, Pellett PL (1994) Plant proteins in relation to human protein and amino acid nutrition. The American Journal of Clinical Nutrition 59:1203S-1212S.

Yu N, Kim M, King ZR, Harris DK, Buck JW, Li Z et al. (2015) Fine mapping of the Asian soybean rust resistance gene Rpp2 from soybean PI 230970. Theoretical and Applied Genetics 128:387-396.

Zhu H, Choi HK, Cook DR, Shoemaker RC (2005) Bridging model and crop legumes through comparative genomics. Plant Physiol 137:1189–1196.

Chapter - 8

Genetic Improvement of *Pongamia Pinnata*: A potential Biodiesel Plant in India

—*Shyam Sunder Sharma, Shashi Bhushan Tripathi*

8.1 Introduction

Pongamia pinnata L. Pierre (also known as Karanja) is a member of family Fabaceae, which has been recognized as a potential source of biodiesel (Tewari 2003). Its seed oil content ranges between 24 per cent and 40 per cent, which is at par with other TBOs like *Jatropha curcas* (Sarma et al. 2005; Kesari et al. 2009a; Kaushik et al. 2007). The seed yield per tree is between 8 to 24 kg (Bringi and Mukerjee 1987; Kesari et al. 2008). However, reports of seed yield up to 100 kg per tree are also available.

Pongamia is believed to be a native of India and south east Asia and has been successfully introduced to humid tropical regions of the world as well as parts of Australia, New Zealand, China and the USA (Scott et al. 2008). In India, Pongamia tree is grown throughout the country except temperate regions. The species occurs naturally in Andhra Pradesh, Chhattisgarh, Gujarat, Orissa, Madhya Pradesh, Uttar Pradesh, Rajasthan, Karnataka, Tami Nadu, Kerala, Jharkhand, Maharashtra, Uttarakhand and West Bengal. Large numbers of trees have been planted on road sides both on highways and in urban areas.

Pongamia is a medium tall tree reaching 10-15 meters. It is an evergreen tree with wide canopy but shades leave in winter under north Indian conditions. The diploid chromosome number (2n) is 22. It grows in all kinds of soils and has high tolerance to drought and salinity but is moderately sensitive to frost and water logging. The leaves are alternate, pinnately compound, green and shiny. Its flower colour varies from lavender to pink to off white. The flowers are complete and papilionaceaous which are arranged in a racemose inflorescence. The pod is smooth, brown, hard and indehiscent at maturity with 1-2 seeds per pod. Mature pods vary from 3-6 cm in length and 2-3 cm in width. Seeds are flat, oval or elliptical or kidney-shaped.

Besides its biodiesel and medicinal properties, Pongamia is also known for its uses as animal fodder, timber and fish poison (Cribb and Cribb 1981). In agriculture, it has been recognized for potential applications due to its antimicrobial, insecticidal, and nematocidal activities (Chopade et al. 2008; Kesari et al. 2009a). As a legume and nitrogen fixing tree, Pongamia has also soil remediation benefits (Scott et al. 2008; Uddin et al. 2009). In a study by (Kesari et al. 2013) Pongamia was found to harbour "rhizobia" as endophytic bacterial community in its root nodules. They isolated a Rhizobium strain, VKLR-01 which exhibited higher tolerance to the prevailing adverse conditions. Such strains can be directly used as inoculants for reclamation of wastelands.

P. pinnata is native to Indian subcontinent. High phenotypic as well as genetic variability has been reported in accessions from India. *Pongamia* showed high phenotypic variation with respect to seed, pod and oil traits. The effectiveness of tree improvement programme mainly depends upon the magnitude and type existing phenotypic variability and also on the degree of transmission of traits in the germplasm (Zobel and Talbert 1984).

It is important to characterize and select candidate plus trees (CPTs) of Pongamia for the improvement of this species in addition to experiments on controlled crossing among selected genotypes (Kesari et al. 2008). A plus tree is an individual tree of a species possessing superior morphological and reproductive characters than other individuals of the same species (Kesari et al. 2010a).

The level of genetic diversity in Indian accessions of Pongamia is expected to be high because of its presumed Indian origin. Scott et al. (2008) also suggested that *P. pinnata* were probably introduced from India to Australia early in human history. High diversity in pod and fruit traits in Pongamia from northern India has been reported earlier (Kaushik et al. 2007). However, very less work has been done on analysis of molecular genetic diversity in Pongamia in comparison to another biofuel crop, Jatropha.

8.2 Energy security and renewable energy

Energy security has emerged as an important policy issue all over the world. India is the fifth largest consumer of primary energy source in the world. Fossil fuels like crude oil, coal and natural gas which, provide almost 80 per cent of the world's primary energy supply, are depleting very fast. Moreover, these sources are non-renewable and contribute to carbon dioxide emissions. The extensive use of these fossil fuels has serious implications on environment particularly on climate change. Therefore, the present emphasis is on alternative sources of energy which are renewable and ecofriendly. This could also reduce our dependence on petroleum fuels. The major renewable energy alternatives are solar, wind, hydroelectric and biofuels. Worldwide, the renewable sources of energy contributed 19 percent to total energy and 22 percent to electricity generation in 2013 with investments in such technologies exceeding US$214 billion in 2013 (2014).

Amongst several alternative renewable sources of energy, biofuels have emerged as a most potent source of energy and several countries and international institutions regard them as "future fuels" as the reservoirs of fossil fuels are declining day by day (Raju et al. 2012). So, it is important to think an alternative and renewable alternative to the fast depleting reserves of fossil fuels (Martini and Shell 1998). It has become apparent that biofuels are destined to impact the future energy security (Scott et al. 2008).

8.3 Biofuels

Biofuels are fuels derived from biomass. Biomass is the organic matter taken from or produced by plants and animals. These include wood, agricultural biomass and products, aquatic plants and animal wastes. The various feed stocks that can be used for the production of biofuels may be classified into cellulosic biomass, sugar and starch producing crops, oils seeds and oil producing plants.

There are two broad categories of biofuels, namely, bioethanol and biodiesel. Bioethanol is used as a replacement for gasoline and biodiesel is used as a replacement for diesel. Biofuels offer a number of environmental, social and economic advantages apart from being a renewable alternative for fossil fuels. The use of biofuels may lead to reduction in vehicular pollution and emission of greenhouse gases as it has been established that the emission of sulphur dioxide, particulate matter and carbon monoxide are less from biofuels (Subramanian et al. 2005).

8.4 Biodiesel

Biodiesel is a renewable, biodegradable fuel that can be which is produced from vegetable oils or animal fats. Biodiesel is prepared by trans esterification of the crude vegetable oil or animal fat with methanol in the presence of NaOH as catalyst with a conversion (oil to biodiesel) of about 90 per cent at 60^0C (Mamilla et al. 2011). Biodiesel is nontoxic, so it would cause less damage than petroleum diesel if spilled or otherwise released to the environment. It is also safer than petroleum diesel because it is less combustible. The flashpoint for biodiesel is higher than 150°C, compared with about 52°C for petroleum diesel. Biodiesel is safe to handle, store, and transport.

8.5 Beginning of biofuel initiatives in India

The Government of India has taken several initiatives for production and use of biofuels during the past decade. The National Mission on Biofuels, which was launched in 2003 envisaged Ethanol Blended Petrol Programme (EBPP) and Biodiesel Blending Programme (BDBP) as its integral components. The feed stocks identified were molasses for the production of ethanol and tree-borne non-edible oilseed crops like Jatropha and Pongamia for the production of biodiesel (Raju et al. 2012).

The 'National Policy on Biofuels' released in 2009 aimed at biofuels as potential means to stimulate rural development and generate employment opportunities. It further aimed to reap environmental and economic benefits arising out of large-scale use of biofuels. The policy set an indicative target of 20 per cent blending of petrol and diesel with bioethanol and biodiesel respectively in the transport sector by the year 2017 (GoI, 2009). However, the challenge was to produce large quantities of biofuels feedstock (seeds) at prices competitive with those of currently used fossil fuel products on a sustainable and environmental friendly basis with particular emphasis on the under-utilized and less productive lands. Further, when most of the cultivable area is occupied by cultivated crops, there is a risk of conflict between food security and biofuels if such agricultural lands are used for biodiesel production. Therefore, the preference is on non-edible oilseeds which can be grown on non-agricultural marginal lands. In this context, non-edible oil seed plants such as Jatropha, Pongamia, Apricot, Mahua etc. have an advantage. The potential of *P. pinnata* as a source of fuel for the biodiesel industry has been well recognized (De and Bhattacharyya 1999; Azam et al. 2005; Karmee and Chadha 2005). Important fuel properties of methyl esters of biodiesel produced from Pongamia oil like viscosity, flash point, fire point, calorific value etc., are within the limits of the standard biodiesel (Kshipra et al. 2013). Therefore, Pongamia can provide biodiesel from non-agricultural lands which is renewable, safe and non-polluting.

8.6 Pongamia as a potential biodiesel crop

Pongamia pinnata L. Pierre (synonym–*Millettia pinnata*) also known as "Karanja", is a member of family Fabaceae, which has been recognized as a major biodiesel species by Planning Commission of India (Tewari 2003). It is currently the most common avenue tree in the cities and towns of India. In states of Karnataka, Andhra Pradesh and Tamil Nadu, it is also found growing on the boundaries of agricultural fields. Besides a potential source of biodiesel, Pongamia has been used as medicine, animal fodder, green manure, timber and fish poison (Meera et al. 2003; Kesari et al. 2008; Sharma and Singh 2008; Kesari et al. 2009a; Scott et al. 2008). As it can be cultivated on marginal land, it has the potential to not only provide a renewable energy resource but in addition will alleviate the competitive situation that exists with food crops as biofuels and associated arable land and water use. This is also in line with the present Biofuel Policy in India. Being a legume, it has also the property of biological nitrogen fixation which can be used in improving the nutritional status of soil. This is why Pongamia is preferred in urban landscaping which also helps in improvement of degraded lands. Considering all these benefits together, Pongamia seems as a good candidate for biodiesel production in India.

8.7 Studies on genetic diversity in Pongamia using molecular markers

Rout et al. (2009) carried out studies on inter and intra-population variability of *Pongamia* using ISSR markers. The study was carried out on a representative population of 111 individuals encompassing seven populations. The molecular polymorphism between individuals was found to be 67.18 per cent with 10 ISSR primers, indicating a modest amount of genetic variation. The within population variation was found to be 32.34 per cent and the polymorphism at species level was found to be 94.34 per cent.

Sujatha et al. (2010) carried out the genetic variability in ten different trees collected from two locations as well as on the in vitro-raised Pongamia plantlets using ISSR markers. They found that, fidelity studies undertaken on the *in vitro*-raised Pongamia plantlets produced no polymorphic bands. This confirmed the uniformity of the clones and validated the reliability of the protocol developed for clonal propagation of this species. On the contrary, high degree of variability (76 per cent) among the selected trees was observed suggesting applicability of the tissue culture techniques for multiplication of high oil-yielding Pongamia trees and exploration of desired wild germplasm characteristics.

Kesari et al.(2010b) used RAPD, ISSR and AFLP markers to assess the genetic variability and relatedness among 10 candidate plus trees (CPTs) of Pongamia from Assam and also to assess the effectiveness of each method. They found polymorphism rates of 10.48 per cent, 10.08 per cent and 100 per cent using 18 RAPD, 12 ISSR and 4 AFLP primer combinations, respectively. They also observed that AFLP had the highest efficiency for estimating the levels of genetic diversity and could be used for fingerprinting, mapping and diversity studies.

Thudi et al. (2010) carried out AFLP analysis of *Pongamia* in order to analyze the genetic diversity amongst 48 accessions collected from 6 different states of India. Five primer combinations generated 520 fragments out of which 502 were polymorphic, giving 96.5 per cent polymorphism. This was one of the confirmations that there is a high level of genetic diversity present in *Pongamia pinnata* across the country. In another study, Kesari et al. (2011) employed RAPD markers to evaluate genetic diversity of 10 candidate plus trees of *P. Pinnata*. The pairwise similarity coefficients in this study ranged from 0.11 to 0.87 which proved presence of high genetic diversity in these genotypes.

Recently, Pavithra et al. (2014) evaluated the genetic diversity of 33 candidates plus trees of *Pongamia pinnata* from five agro-ecological zones of southern Peninsular India using AFLP markers. Cluster analysis identified three major clusters and four subgroups depending on their geographical location. This study revealed relatively higher levels of gene diversity and higher number of unique bands in accessions from eastern dry zone of Karnataka and southern dry and transition zone of Karnataka.

Sharma et al. (2011) analyzed the genetic diversity of 20 *Pongamia* accessions using two dominant molecular markers namely, AFLP and TE-AFLP. They also found that, values of effective multiplex ratio, marker index, and resolving power were markedly higher in AFLP than in TE-AFLP. Inferences based on TE-AFLP data had similar level of biological relevance as compared to AFLP data when location and diameter of trees were taken in to consideration. However, TE- AFLP was easier to score than those obtained by AFLP, making TE-AFLP the method of choice to be used when large numbers of accessions need to be genotyped.

In a study by Sharma et al. 2016, a total of 275 Pongamia individuals from different locality of NCT of Delhi genotyped using six TE-AFLP primer combinations. Pongamia exhibited very high genetic diversity is present even within the accessions from Delhi and surrounding regions. All accessions grouped into six major clusters and highest genetic diversity was found in the accessions from South Delhi. It may be noted that this area of National Capital harbours the largest number of Pongamia trees. High diversity in pod and fruit traits in Pongamia from a neighboring area of NCT of Delhi has been reported earlier (Kaushik et al. 2007). A high level of genetic diversity in Pongamia is expected because of its presumed Indian origin. Scott et al. (2008) also suggested that *P. pinnata* were probably introduced from India to Australia early in human history.

It may be noted that the Department of Horticulture, is responsible for these plantations, procures seeds from different sources across India to raise planting material for its avenue plantations. In many instances, trees with similar girth at breast height (GBH) values within a locality were found to have relatively higher genetic similarity and therefore, grouped together. It may be due to the fact that trees with similar GBH within a particular locality might have been planted in the same year and may belong to the same seed stock. On the other hand, trees with significantly different GBH are most likely plantations of different years. Out of the 20 core accessions identified in this by Sharma et al. (2016b), 9 were from the South Delhi region itself. This finding is in congruence with the findings of genetic diversity analysis and AMOVA which showed that there was highest genetic diversity within South Delhi accessions.

A major outcome of the study is the information on the extent and distribution of genetic diversity in various populations of Pongamia, which can be utilized in its genetic improvement based on breeding and selection. These accessions can be used for new plantations by utilizing clonal techniques such as macro and micropropagation. As a long term strategy, the diverse accessions can be used to generate new and improved genotypes through controlled crossing and field selection. Several markers associated with oil content were detected in this study. Some of these with highest contribution could be used to develop markers for application in breeding programs through marker assisted selection.

A number of studies have been conducted on Pongamia using dominant markers such as RAPD, ISSR, AFLP and TE-AFLP (Sharma et al. 2011; Sahoo et al. 2010; Kesari et al. 2010b). However, the information content of these dominant markers is less due to presence of only two allelic states. Codominant markers such as microsatellites can display high number of allelic states thus providing higher information per data point. However, these markers need to be developed in the species of interest through generating sequence information and primer development.

To best of our knowledge, no microsatellite markers have been reported in Pongamia till date. The aim of the present study was to isolate and characterize polymorphic microsatellite markers as to facilitate genetic diversity, clonal identification and conservation in *P. pinnata* germplasm. A total of thirty nine microsatellite containing sequences were isolated and submitted to the GenBank with accession number GQ130157 to GQ130195 (Sharma et al. 2014). Nineteen of these sequences had repeats with enough flanking nucleotides were used for primer designing and clone sequences of nine polymorphic microsatellite loci. Nine polymorphic microsatellite loci identified in this study showed a clear and single peak for each allele while remaining 10 markers showed no polymorphism across accessions. The amplified products were in agreement with the expected size from sequence data for all nine microsatellites analyzed. Hardy Weinberg and Linkage disequilibrium tests were performed and all nine polymorphic loci were in HWE ($P < 0.005$) and significant linkage disequilibrium between pairs of loci were found which shows that the loci were not linked. The PIC value obtained in this study was significantly higher than the earlier study done on Pongamia using dominant AFLP and TE-AFLP markers (Sharma et al. 2011).

The analysis of genetic diversity using nine polymorphic microsatellites revealed a reasonably high level of diversity in 24 accessions used in this study (Sharma et al. 2014). The results of this study are in congruence with our earlier study on Pongamia using AFLP and TE-AFLP markers (Sharma et al. 2011). The biased grouping of accessions with high oil content in cluster 1 and accessions with low oil content in cluster 2 may be due to random association of some of these loci with oil content. A much larger number of accessions need to be studied to obtain statistically valid associations between microsatellites and oil content. Despite the modest number of accessions analyzed in this study, the results indicate that the microsatellite loci developed in this study will serve as very useful tool for genetic diversity analysis, clonal identification and conservation in *P. pinnata* germplasm.

8.8 Oil content and fatty acid analysis in Pongamia

Pod development and seed maturation in Pongamia took 8 to 9 months after anthesis. A similar duration has been reported earlier in Pongamia (Pavithra et al. 2012). Significant changes in oil content and fatty acid composition occurred during this period. Oil content increased from 16.25 per cent at stage 1 to 36.83 per cent at stage 8. A discrepancy in this observation is noted when compared to results of Pavithra and coworkers (2012) who reported an increase from 32 to 36.53 per cent during seed development. The discrepancies in findings of the current study and Pavithra and coworkers (2012) is likely to be due to different maturity stages which were sampled in the two studies. The stages sampled by Pavithra and coworkers (2012)coincide with the Stages 6 to 8 of this study besides some additional earlier stages. Increase in oil content during seed development and growth has also been reported in Jatropha curcas, and algae (Dhup and Dhawan 2013; Annarao et al. 2008).

The range of variations in all fatty acids during seed development in this study was much higher than reported earlier (Pavithra et al. 2012). Once again, these discrepancies can be accounted for by considering different stages sampled in the two studies. The levels of palmitic acid and linoleic acid in mature seeds in both studies were comparable. However, the oleic acid content observed in the current study was much higher than that reported in the earlier study (Pavithra et al. 2012). The variation in oleic acid content and absence of erucic acid in mature seeds of the present accession, P015, may be due to genetic and/or environmental factors. Considering high genetic and phenotypic variability observed in Pongamia from India, effect of genetic factors on seed oil composition is very likely (Sharma et al. 2011; Kaushik et al. 2007).

The study of seed traits and oil content of the natural populations is useful in exploiting the genetic variation, as they are important component of yield (Kaushik et al. 2007; Kesari et al. 2008). Seed weight showed highly significant correlation with seed length, seed breadth and seed thickness. This is expected as all the three parameters are responsible for determining seed size. The oil content showed a positive and significant correlation with all seed characters studied. Oil yield is one of the most important traits for commercial success of *P. pinnata* as an energy crop. In the present study oil content was highly variable across 157 accessions. It ranged from 23.5 to 40 per cent. The results are similar to those reported by Kaushik et al. (2007). A much higher range of 9.5 to 46 per cent has been reported in accessions from Andhra Pradesh (Mukta et al. 2009). Similarly, another study on 123 accessions from Andhra Pradesh and Orissa showed oil content 15 to 47 per cent (Sunil et al. 2009).

The fatty acid composition of oil governs its chemical properties and its efficacy as biodiesel. The predominant fatty acids in Pongamia are oleic acid, palmitic acid, stearic acid and linoleic acid. Fatty acid composition of Pongamia is suitable for biodiesel according to USA, German and European standards (Azam et al. 2005; Karmee and Chadha 2005). Oleic acid is the most desirable fatty acid for good quality biodiesel, due to its low viscosity and low melting point, which makes it suitable for use in cold climates (Knothe 2009). However, other major fatty acids, palmitic and stearic and linoleic posseses good biodiesel properties. The oleic acid content observed in present accessions is slightly higher than the previous report (Mukta et al. 2009). Bala and coworkers (Bala et al. 2011) reported the presence of significant levels of erucic acid in Pongamia trees from Western India. The absence of erucic acid in the present accessions may be due to genetic as well as environmental factors.

In a study clustering analysis of accessions based on fatty acid composition grouped all the accessions in three major clusters (Sharma et al. 2016a). Cluster 1 contained accessions with both low as well as high oleic acid content while clusters 2 and 3 grouped the accessions with high and low oleic acid content respectively. In an earlier study, very high similarity was observed among Indonesian genotypes while

Java type was quite distinct from all other genotypes based on fatty acid profiles (Arpiwi et al. 2012).

The values of biodiesel traits such as SN, IV and CN of Pongamia accessions were empirically calculated. The calculated SN and CN values are in good agreement with previously determined values by Mukta and coworkers (2009). The degree of unsaturation, measured as IV, is also an important criterion for selection of suitability of biodiesel. The presence of unsaturated fatty acids to an extent is important as it prevents the solidification of oil but after a certain degree of unsaturation, the oil is not suitable for biodiesel as it polymerizes in to plastic like substance. For this reason, biodiesel standards require that the IV should be not be more than the specified value. In our study, all the accessions analyzed had IV less than 115 meeting the requirement.

CN is an important parameter, which is considered during the selection of FAMEs for use as biodiesel and determines the ability of fuel to ignite quickly after being injected. Higher the CN value better is the ignition quality of a fuel. Biodiesel standards of USA, Germany and European organization have specified minimum value as 47, 49 and 51 respectively. In the current analysis of 38 Pongamia accessions, five accessions namely, P008, P070, P165, P203 and P213 had CN values more than 55 which were classified as high CN value accessions (Sharma et al. 2016a).

8.9 Estimation of outcrossing rates in Pongamia

Estimation of outcrossing rates is important to develop breeding and conservation strategies. The, an effort was made to estimate outcrossing rates in Pongamia using AFLP and SSR markers. The estimates of multilocus (t_m) and single-locus (t_s) outcrossing rates obtained with MLTR from both AFLP and SSR data, clearly indicated that *Pongamia* is a predominantly an outcrossed species (Sharma et al. 2017). The multilocus outcrossing rate estimates based on all 125 AFLP loci and 28 SSR alleles were 91.7 per cent and 74.8 per cent respectively. A higher multilocus and single locus estimate with AFLP markers as compared to microsatellite markers may be due to their large number which enables more efficient detection of outcrossing events. Multilocus estimates significantly differed from the mean single-locus estimates for both the markers. The difference between tm and ts indicated that about 20 per cent of the outcrosses occurred due to mating between relatives (biparental inbreeding). The estimated value of Fixation index, F in the progeny using both AFLP and microsatellite data was 0.198 and 0.181 respectively. On the other hand, the expected F for AFLP and microsatellite markers and formula [F = (1 - t) / (1 + t)] was 0.0432 and 0.144 respectively (Sharma et al. 2017). The estimated F was higher than expected F with both the markers which suggests a loss of heterozygotes and higher inbreeding in the progeny than expected under equilibrium.

Previous studies have used both dominant AFLP markers (Gaiotto et al. 1997; Muluvi et al. 1999) and codominant microsatellite markers (Connell et al. 2004; James et al. 2008; Hon-xing et al. 2008) to estimate the outcrossing rates in many tree species. AFLP markers show dominance with lower information content per locus than co-dominant microsatellites. However, the high multiplex ratio of AFLP markers allows the generation of sufficient markers to obtain a robust estimate of outcrossing rates. No study has so far reported in Pongamia on outcrossing rates using molecular markers. An earlier report based on study of nectar-feeding activities of certain bees and on explosive keel mechanism showed that these conditions promote cross pollination and discourages selfing in Pongamia (Raju and Rao 2006). The results of the present work also support this hypothesis and demonstrate the usefulness of AFLP and microsatellite markers in studying the mating-system in *Pongamia*.

8.10 Population and association analysis in Pongamia

Population structure was detected in the dataset using STRUCTURE. The Bayesian method identified the optimal number of clusters based on allele frequencies. Two indicators for identification of optimum number of populations (K) were used in this study. In the first approach, the optimum number of K was identified directly from the maximum likelihood, LnP(D), as described by Pritchard et al. (2000). However, this approach did not show any clear indication of number of populations. Evanno et al. (2005) reported that the estimated number of populations (K) based on LnP(D) are not correct in many cases. They recommended an *ad hoc* quantity called delta K ("K) based on the rate of change in the log probability of data between successive K values provided by STRUCTURE. In the present study, although LnP(D) was not able to clearly detect the number of population, "K gave a clear peak at K=9. Considering the diverse sources of accessions, this number is very likely. It may be noted that most of the Pongamia plantations are carried out by state agencies of horticulture and forestry, who procure seeds from different sources. The sources may also change year to year. Therefore, it is likely that a relatively smaller geographical area may harbor accessions of different origin. In many instances, trees with similar girth at breast height (GBH) values within a locality were found to have relatively higher genetic similarity and therefore grouped together. It may be due to the fact that trees with similar GBH within a particular locality might have been planted in the same year and may belong to the same seed stock. On the other hand, trees with significantly different GBH are most likely plantations of different years.

In a study (Sharma et al. 2016b)described association analysis of seed oil content with TE-AFLP markers identified two markers, namely, E_ac x P_a39 and E_ac x P_a04, to be associated with oil content with very high statistical significance (P=0.005). Additionally, 7 and 9 markers were associated with oil content at 99 per cent and 95 per cent confidence level. It may be noted that oil content is supposed to be a polygenic trait and therefore several quantitative trait loci may be involved in its regulation. Moreover, many of the markers may be linked with each other which may be the reason behind large number of loci found in linkage disequilibrium with oil content. Unfortunately, no linkage map is available in Pongamia to substantiate these possibilities. However, the markers, E_ac x P_a39 and E_ac x P_a04, need to be validated on an independent dataset for developing markers for marker assisted breeding.

8.11 Conclusion

Availability of good quality feed stock is important for the success of any biodiesel initiative. Most of the existing trees of Pongamia were planted to develop green belts keeping its aesthetic value. Consequently, a large proportion of these trees have poor seed productivity. Selection and genetic improvement of Pongamia genotypes with superior yields is therefore, required. At the time when this work was initiated, not much information on molecular genetic diversity of Pongamia was available. However, the work on phenotypic diversity related to seed, pod and oil quality traits were available. The information of genetic diversity detected in Pongamia accessions in this study can be used to initiate programs on genetic improvement based on controlled crosses and selection. Some of the accessions identified have several desirable traits which can be directly used for progeny and clonal level evaluation. The newly developed molecular tools, especially the microsatellite markers, can be used for germplasm conservation and management. Long-term programs are required to be initiated in areas such as linkage and QTL mapping for important traits. This should be ideally done by the national level institutes which are working in the area of forestry as they have the required resources as well as the research mandate for this kind of activities.

References

Annarao S, Sidhu OP, Roy R, Tuli R, Khetrapal CL (2008) Lipid profiling of developing *Jatropha curcas* L. seeds using (1)H NMR spectroscopy. Bioresource Technology 99:9032-9035.

Arpiwi NL, Yan G, Barbour EL, Plummer JA (2012) Genetic diversity, seed traits and salinity tolerance of *Millettia pinnata* (L.) Panigrahi, a biodiesel tree. Genetic Resources and Crop Evolution doi:10.1007/s10722-10012-19866-y.

Azam MM, Waris A, Nahar NM (2005) Prospects and potential of fatty acid methyl esters of some non-traditional seed oils for use as biodiesel in India. Biomass and Bioenergy 29:293-302.

Bala M, Nag TN, Kumar S, Vyas M, Kumar A, Bhogal NS (2011) Proximate composition and fatty acid profile of *Pongamia pinnata*, a potential biodiesel crop. Journal of the American Oil Chemists' Society 88:559-562.

Bringi NV, Mukerjee SK (1987) Non-Traditional Oilseeds and Oils of India. In NV Bringi, ed, Oxford IBH Publishing Co. Pvt. Ltd., New Delhi pp 143-166.

Chopade VV, Tanker AN, Pande VV, Tekade AR, Gowekar NM, Bhandari SR et al. (2008) *Pongamia pinnata*: phytochemical constituents, traditional uses and pharmacological properties: a review. International Journal of Green Pharmacy 2:72-75.

Connell LMO, Russell J, Ritland K (2004) Fine-scale estimation of outcrossing in western red cedar with microsatellite assay of bulked DNA.Heredity 1-7

Cribb AB, Cribb JW (1981) Useful Wild Plants in Australia. Collins, Sydney.

De BK, Bhattacharyya DK (1999) Biodiesel from minor vegetable oils like karanja oil and nahor oil. Lipid Fett 101:404-406.

Dhup S, Dhawan V (2013) Effect of nitrogen concentration on lipid productivity and fatty acid composition of *Monoraphidium* sp. Bioresource Technology http://dx.doi.org/10.1016/j.biortech. 2013. 1011.1068.

Evanno G, Regnout S, Goudet J (2005) Detecting the number of clusters of individuals using the software STRUCTURE: a simulation study. Molecular Ecology 14:2611-2620.

Gaiotto FA, Bramucci M, Grattapaglia D (1997) Estimation of outcrossing rate in a breeding population of *Eucalyptus urophylla* with dominant RAPD and AFLP markers. Theor Appl Genet 95:842-849.

Hon-xing RAO, Patterson B, Potts B, Vaillancourt R (2008) A microsatellite study on outcrossing rates and contamination in an Eucalyptus globules breeding arboretum. Journal of Forestry Research 19:136-140.

James W, Borrone, Cecile T, Olano D, avid N, Kuhn J et al. (2008) Outcrossing in Florida Avocados as measured using microsatellite markers. Journal of the American Society for Horticultural Science 133:255-261.

Karmee SK, Chadha A (2005) Preparation of biodiesel from crude oil of *Pongamia pinnata*. Bioresource Technology 96:1425-1429.

Kaushik N, Kumar S, Kumar K, Beniwal R, Kaushik N, Roy S (2007) Genetic variability and association studies in pod and seed traits of *Pongamia pinnata* (L.) Pierre in Haryana, India. Genetic Resources and Crop Evolution 54:1827-1832.

Kesari V, Das A, Rangan L (2009a) Physico-chemical characterization and microbial assay from seed oil of *Pongamia pinnata*, potential biofuel crop. Biomass Bioenergy 34:108-115.

Kesari V, Krishnamachari A, Rangan L (2008) Systematic characterisation and seed oil analysis in candidate plus trees of biodiesel plant, Pongamia pinnata. Annals of Applied Biology 152:397-404.

Kesari V, Madurai V, Sathyanarayana, Parida A, Rangan L (2010a) Molecular marker-based characterization in candidate plus trees of *Pongamia pinnata*, a potential biodiesel legume. Annals of Botany Plants doi: 10.1093/aobpla/plq1017.

Kesari V, Madurai V, Sathyanarayana, Parida A, Rangan L (2010b) Molecular marker-based characterization in candidate plus trees of *Pongamia pinnata*, a potential biodiesel legume. AoB Plants 10.1093/aobpla/plq017:

Kesari V, Ramesh AM, Rangan L (2013) Rhizobium pongamiae sp. nov. from Root Nodules of *Pongamia pinnata*. BioMed Research International 2013:9.

Kesari V, Rangan L (2011) Genetic diversity analysis by RAPD markers in candodate plus trees of *Pongamia pinnata*, a promising source of bioenergy. Biomass Bioenergy 35:3123-3128

Knothe G (2009) "Designer" biodiesel: optimizing fatty ester composition to improve fuel properties. Energy Fuels 22:1358-1364.

Kshipra G, Gupta NC, Sharma DK (2013) Physical characterization and comparison of biodiesel produced from edible and non-edible oils of *Madhuca indica* (mahua), *Pongamia pinnata* (karanja), and *Sesamum indicum* (til) plant oilseeds. Biomass Conversion and Biorefinery DOI 10.1007/s13399-013-0101-7:

Mamilla VR, Mallikarjun MV, Rao GLN (2011) Preparation of Biodiesel from Karanja Oil. International Journal of Energy Engineering 94-100.

Martini N, Shell JS. 1998. Plant oils as fuels- present state of science and future development. Berlin: Springer. p. 276.

Meera B, Kumar S, Kalidhar SB (2003) A review of the chemistry and biological activity of *Pongamia pinnata*. Journal of Medicinal and Aromatic Plant Science 25:441-465.

Mukta N, Murthy IYLN, Sripal P (2009) Variability assessment in *Pongamia pinnata* (L.) Pierre germplasm for biodiesel traits. Industrial Crops and Products 29:536-540.

Muluvi GM, Sprent JI, N S, Provan J, Odee D, Folkard G et al. (1999) Amplified fragment length polymorphism (AFLP) analysis of genetic variation in *Moringa oleifera* Lam. Mol Ecol 8:463-470.

Pavithra HR, Gowda B, Rajesh Kumar K, Prasanna KT, Shivanna MB (2012) Oil, Fatty Acid Profile and Karanjin Content in Developing *Pongamia pinnata* (L.) Pierre Seeds. Journal of the American Oil Chemists' Society 89:2237-2244.

Pavithra HR, Shivanna MB, Chandrika K, Prasanna KT, Gowda B (2014) Genetic analysis of Pongamia pinnata (L.) Pierre populations using AFLP markers. Tree Genetics & Genomes 10:173-188.

Pritchard J, Stephens M, Donnelly P (2000) Inference of population structure using multilocus genotype data. Genetics 155:945- 959.

Raju AJS, Rao SP (2006) Explosive pollen release and pollination as a function of nectar feeding activity of certain bees in the biodiesel plant, *Pongamia pinnata* (L.) Pierre (Fabaceae). Curr Sci 90:960-967.

Raju SS, Parappurathu S, Chand R, Joshi PK, Kumar P, Msangi S (2012) Biofuels in India: Potential, Policy and Emerging Paradigms, Policy paper 27. National Centre for Agricultural Economics and Policy Research, New Delhi.

REN21 (2014) Renewables 2014 Global Status Report (Paris: REN21 Secretariat, 2014). http://www.ren21.net/portals/0/documents/resources/gsr/2014/gsr2014_full%20report_low%20res.pdf.

Rout GR, Sahoo DP, Aparajita S (2009) Studies on Inter and intra-population variability of *Pongamia pinnata*: a bioenergy legume tree. Crop Breeding and Applied Biotechnology 9:268-273.

Sahoo D, Aparajita S, Rout G (2010) Inter and intra-population variability of *Pongamia pinnata*: a bioenergy legume tree. Plant Systematics and Evolution 285:121-125.

Sarma AK, Konwer D, Bordoloi PK (2005) A comprehensive analysis of fuel properties of biodiesel from koroch seed oil. Energy Fuels 19:656-657.

Scott PT, Pregelj L, Chen N, Hadler JS, Djordjevic MA, Gresshoff PM (2008) Pongamia pinnata: An untapped resource for the biofuels industry of the future. Bioenergy Research 1:2-11.

Sharma SS, Islam MA, Malik AA, Kumar K, Negi MS, Tripathi SB (2016a) Seed traits, fatty acid profile and genetic diversity assessment in *Pongamia pinnata* (L.) Pierre germplasm. Physiol Mol Bio Plants 22:193-205.

Sharma SS, Islam MA, Negi MS, Tripathi SB (2014) Isolation and characterization of a first set of nine polymorphic microsatellite loci in *Pongamia pinnata* (Fabaceae). Journal of Genetics 93:e70-74.

Sharma SS, Islam MA, Negi MS, Tripathi SB (2017) Estimation of outcrossing rates in biodiesel species *Pongamia pinnata* based on AFLP and microsatellite markers. National Academy Science Letters 40:105-108.

Sharma SS, Islam MA, Singh VK, Negi MS, Tripathi SB (2016b) Genetic diversity, population structure and association study using TE-AFLP markers in Pongamia pinnata (L.) Pierre germplasm. Tree Genetics & Genomes 13:6

Sharma SS, Negi MS, Sinha P, Kumar K, Tripathi SB (2011) Assessment of genetic diversity of biodiesel species *Pongamia pinnata* accessions using AFLP and three endonuclease -AFLP. Plant Mol Biol Rep 29:12-18.

Sharma YC, Singh B (2008) Development of biodiesel from karanja, a tree found in rural India Fuel 87:1740-1742.

Subramanian KA, Singal SK, Saxena M, Singhal S (2005) Utilization of liquid biofuels in automobile diesel engines: An Indian perspective. Biomass and Bioenergy 29:65-72.

Sujatha K, Rajwade AV, Gupta VS, Hazre S (2010) Assessment of *Pongamia pinnata* (L.) – a biodiesel producing tree species using ISSR markers. Curr Sci 99:1327-1329.

Sunil N, Kumar V, Sivaraj N, Lavanya C, Prasad RBN, Rao BVSK et al. (2009) Variability and divergence in *Pongamia pinnata* (L.) Pierre germplasm-a candidate tree for biodiesel. GCB Bioenergy 1:382-391.

Tewari DN (2003) Report of the committee on development of Bio-fuel. Planning Commission, Government of India.

Thudi M, Revathi M, Wani SP, Tatikonda L, Hoisington DA, Varshney RK (2010) Analysis of genetic diversity in Pongamia [*Pongamia pinnata* (L) Pierre] using AFLP markers. Journal of Plant Biochemistry and Biotechnology 19:209-216.

Uddin MB, Mukul SA, Khan MASA, Hossain KM (2009) Seedling response of three agroforestry tree species to phos phorous fertilizer application in

Bangladesh: growth and nodulation capabilities. Journal of Forest Research 20:45-48.

Zobel BJ, Talbert J (1984) Applied forest tree improvement. John Wiley and Sons. New York, 505.

Index